Key Topics in Food
Science and Technology – No. 3

Food Manufacturing: An Overview

Tim Hutton

Campden & Chorleywood Food Research Association Group comprises
Campden & Chorleywood Food Research Association
and its subsidiary companies
CCFRA Technology Ltd   CCFRA Group Services Ltd   Campden & Chorleywood Magyarország

Campden & Chorleywood Food
Research Association Group

Chipping Campden, Gloucestershire, GL55 6LD UK
Tel: +44 (0) 1386 842000 Fax: +44 (0) 1386 842100
www.campden.co.uk

Information emanating from this company is given after the exercise of all reasonable care and skill in its compilation, preparation and issue, but is provided without liability in its application and use.

Legislation changes frequently. It is essential to confirm that legislation cited in this publication and current at the time of printing is still in force before acting upon it. Any mention of specific products, companies or trademarks is for illustrative purposes only and does not imply endorsement by CCFRA.

ISBN: 0 905942 35 3

SERIES PREFACE

Food and food production have never had a higher profile, with food-related issues featuring in newspapers or on TV and radio almost every day. At the same time, educational opportunities related to food have never been greater. Food technology is taught in schools, as a subject in its own right, and there is a variety of food-related courses in colleges and universities - from food science and technology through nutrition and dietetics to catering and hospitality management.

Despite this attention, there is widespread misunderstanding of food - about what it is, about where it comes from, about how it is produced, and about its role in our lives. One reason for this, perhaps, is that the consumer has become distanced from the food production system as it has become much more sophisticated in response to the developing market for choice and convenience. Whilst other initiatives are addressing the issue of consumer awareness, feedback from the food industry itself and from the educational sector has highlighted the need for short focused overviews of specific aspects of food science and technology with an emphasis on industrial relevance.

The *Key Topics in Food Science and Technology* series of short books therefore sets out to describe some fundamentals of food and food production and, in addressing a specific topic, each issue emphasises the principles and illustrates their application through industrial examples. Although aimed primarily at food industry recruits and trainees, the series will also be of interest to those interested in a career in the food industry, food science and technology students, food technology teachers, trainee enforcement officers and, established personnel within industry seeking a broad overview of particular topics.

Leighton Jones
Series Editor

PREFACE TO THIS VOLUME

The relative abundance of food in the developed world means that attitudes to it have changed markedly over the last 50 years or so. As it has contributed to growing material comfort, food consumption has become regarded not just as a basic necessity but as a recreational activity. The proliferation in choice facing the food shopper has necessarily been paralleled by the increasing sophistication of the food supply chain, with industrialisation of food production and preservation an inextricable part of the general development of the industrial world.

For many foods, the hygienically packed end-product, attractively presented in the modern food store, almost belies the 'crude' biological origin of the raw materials. It also belies the extensive effort required to make the food safe and palatable and to prevent (or at least slow down) its inevitable (and entirely natural) biological and chemical deterioration.

This short book presents an overview of food manufacturing in the context of the modern food supply chain and the changing marketplace. It describes what food is, highlights some of the technical constraints faced by food companies and describes how these sit alongside technological innovation and the long established principles of food preservation. It is intended, however, only as a brief introduction - as a lead-in to the many detailed texts that already exist. It is not intended as an in-depth review of specific technologies nor as a discussion of past or emerging 'food scares'. Rather, it is a collection of snapshots that illustrate some common principles and practices.

Tim Hutton
CCFRA

ACKNOWLEDGEMENTS

In compiling this guide I am grateful to the following of my colleagues for their constructive comments and advice: Peter Coggins, Brian Day, Prof. Colin Dennis, Margaret Everitt, Martin Hall, John Hammond, John Holah, Dr. Leighton Jones, Dr. Mike Stringer, Dr. Steven Walker, Celia Willcox and Dr. Alan Williams. Some of the examples used to illustrate specific points are drawn from other CCFRA publications and I am grateful to the authors of these (as cited) for providing the source material. Finally thanks are also due to Janette Stewart for the artwork and design.

NOTE

All definitions, legislation, codes of practice and guidelines mentioned in this publication are included for the purposes of illustration only and relate to UK practice unless otherwise stated.

CONTENTS

1. INTRODUCTION – FOOD IN PERSPECTIVE

The food industry is now a highly complex, multi-billion pound, international industry. From the consumer's point of view, the industry in the UK is dominated by the major supermarkets. It is easy to take for granted the abundance of choice that they and other independent retailers offer, with fresh produce sourced from all over the world, and around 15-20,000 product lines in total in a standard out-of-town store. These are the end products of a global production, processing and distribution system; for example, a ready meal assembled in the UK may contain ingredients from mainland Europe, Asia and the Americas. As well as developing a network that is effective, the food industry, like all industries, has to be policed to ensure that legislative standards and guidelines are being met and that the consumer is being fairly treated.

It is also easy to forget that food is biological and a product of the natural world. Consequently, many food materials can vary considerably in their composition, quality attributes, processing properties and nutritional quality. For example, the way in which a particular wheat or potato crop can and will be used may depend on: the crop variety; the location in which it was grown; the weather conditions during growth and harvest; the use of particular agronomic practices (e.g. irrigation, application of fertilizers); post-harvest storage conditions and duration; the presence of natural pests and diseases during growth and storage and the measures taken to protect the crop; and so on. All of these variables and the biological nature of foods mean that many raw materials may naturally contain contaminants - physical, chemical and microbiological. These have to be reduced to a minimum before the food is deemed fit for consumption, even if the end material is still a raw, unprocessed fruit or vegetable. The end product also has to be acceptable in terms of sensory quality (e.g. flavour, colour, texture) as well as price, and it has to be delivered to outlets that are convenient for the consumer to visit.

1.1 Brief history of food production

The food industry effectively began when mankind changed from a hunter-gatherer lifestyle to a farming and trading lifestyle, although trading in and bartering with food has probably occurred throughout the history of mankind. The growing and processing of cereal and vegetable crops and the keeping of livestock and production of meat, dairy and poultry products were among the first activities of the new farming and trading industry.

Even before farming and agricultural trading began in any formal way, people were inventive about what they did with the food they gathered. Berries were fermented into alcoholic drinks, meat and vegetables were cooked, and meat which could not be eaten after a kill was stored (either before or after cooking) by such methods as freezing, smoking or salting. (Freezing was carried out in the northern parts of Europe by burying the food in the permafrost.) In hot climates, some raw materials could be left to the effects of natural microbial fermentation to produce a relatively stable, and safe, end-product. Thus, the processing of food is not specifically a modern-day development.

The advent of a farming culture accelerated the development and increased the scale of production of dairy products such as cheese and yoghurt, and of cereal-based products like bread. Different cultures developed a variety of processed foods depending on the natural resources available to them, the amount of trade they could carry out, and the environmental conditions where they lived. In 450 A.D., Attila the Hun reportedly preserved meat by storing it under his saddle, where sweat from his horse provided ideal preservation and processing conditions. History doesn't relate what it tasted like!

Until the 19th century, long-term storage of food generally relied on processing to yield a product which was inherently ambient-stable. This meant the development of various cured, salted, pickled, fermented and smoked products, as well as jams and other high-sugar products, and a variety of dried foods. Apart from jams and foods with very low amounts of water, the shelf-lives of most of these products would be measured in weeks rather than years.

The shelf-life of the more perishable products was extended by keeping the food in cool larders. This method was in common use in the UK until the 1960s. Physical barriers such as fly screens were also used to keep out flies and similar pests. It was in 1665 that Francesco Redi, an Italian physician, showed that the development of maggots in meat was not a spontaneous occurrence, but in fact was caused by flies being able to deposit their eggs in the meat. Protecting the meat with fine gauze prevented the entry of the flies and the growth of the larval stage - the maggot (Stanier *et al.*, 1976).

The first technology which could be applied across the whole spectrum of food types to give a shelf stable end product was the canning of heat processed foods in a hermetically sealed container. This has its origins with Nicholas Appert around 1810 and Napoleon Bonaparte was an early purchaser of canned foods for his troops. However, it was not until the 1920s that the canning industry was firmly established in the UK, although it had been well established elsewhere (especially in the USA) before this. (For further reading on the history of food preservation, see Thorne, 1986 and Shepherd, 2000).

The advent of the refrigerator in the 1950s meant that perishable food previously stored in cool larders could now be kept more conveniently (and more safely). The growth in fridge ownership was very rapid - in 1955 only 8% of UK households had them, but by the early 1960s they were present in nearly all households. They usually had a small freezer box for storing ice-cream and/or frozen peas and burgers, but the variety of frozen food available was limited.

Frozen food technology in the UK had been developed in the 1950s, but not until the 1970s did the ownership of stand-alone freezers become widespread and the huge array of frozen foods, including ready meals, become available. This was followed by the development of chilled ready meals, in the race to provide more convenient food of improved quality.

The most recent technological development to have a major impact on the food industry from a consumer's point of view has been the microwave oven. In 1985, 18% of UK households possessed a microwave, but by 1996 this had risen to around 70%. As a result, through the 1990s 13-18% of new products introduced to the UK marketplace had microwave cooking instructions (which equates to the majority of new products that actually need cooking or reheating).

As well as these major developments, there have been many product-specific developments such as the pasteurisation of milk in the 1920s, the availability of instant coffee (initially introduced by Nestlé in 1938) and the production of tea in tea bags by Tetley in 1952. Some of the developments in products and processes, with approximate dates, are given in Tables 1 and 2.

Table 1 - Some landmarks in product development

1869	Margarine developed
1872	Chewing gum introduced
1886	Coca-Cola introduced
1889	First Hamburger
1930	Quick-frozen peas
1932	Mars bar
1936	Canned beer produced
1937	Smarties
1938	Nestlé instant coffee
1943	Lucozade first aids recovery
1948	First holes in polo mints
1952	Tetley put tea in bags
1970s	Chilled ready meals
1996	Tomato paste from GM tomatoes

By far the biggest trade development to have influenced the food industry in the UK in the last 40 years has been the growth of the major supermarket retailers, and they have been a major driving force behind the development of many of the vast array of formulated foods now available. They have also influenced the structure, standards and economics of the food supply chain to a considerable extent.

Table 2 - Landmarks in process and machinery development

1809	Canning first developed
1834	The first refrigerator
1842	Freezing process patented
1850	Dish washers developed
1899	High-pressure processing developed
1929	Freezing process commercialised
1940	Freeze drying
1947	First microwave oven
1954	The Tefal pan is introduced
1961	Chorleywood bread process
1965	UHT milk
1968	Food irradiation becomes possible
1975	High-pressure process commercialised

1.2 Plant and animal domestication

Most of what we eat today is grown or farmed specifically for that purpose. With the exception of fish and other seafood, we eat very little from the wild. In the UK, around 76% of the total land area is devoted to agriculture of some sort. Figure 1 gives a breakdown of the area devoted to arable farming and horticulture in the UK.

Domestication of crops and livestock has gone hand-in-hand with the development of a secure food supply and the emergence of industrial food production. The extent to which domestication has changed crops and livestock, and the way in which geographical redistribution has occurred are generally not widely appreciated. The wild ('natural') forms of potato, tomato and celery are all poisonous; carrots were originally white, not orange; broccoli, Brussels sprouts, cabbage, kale and cauliflower all derive from the same wild species (*Brassica oleracea*); and there are many very different varieties of beans derived from the original wild *Phaseolus vulgaris* (see Bedford, 1986). Cereals have also been extensively bred so that domestic wheat is not diploid (i.e. with two sets of chromosomes), like most

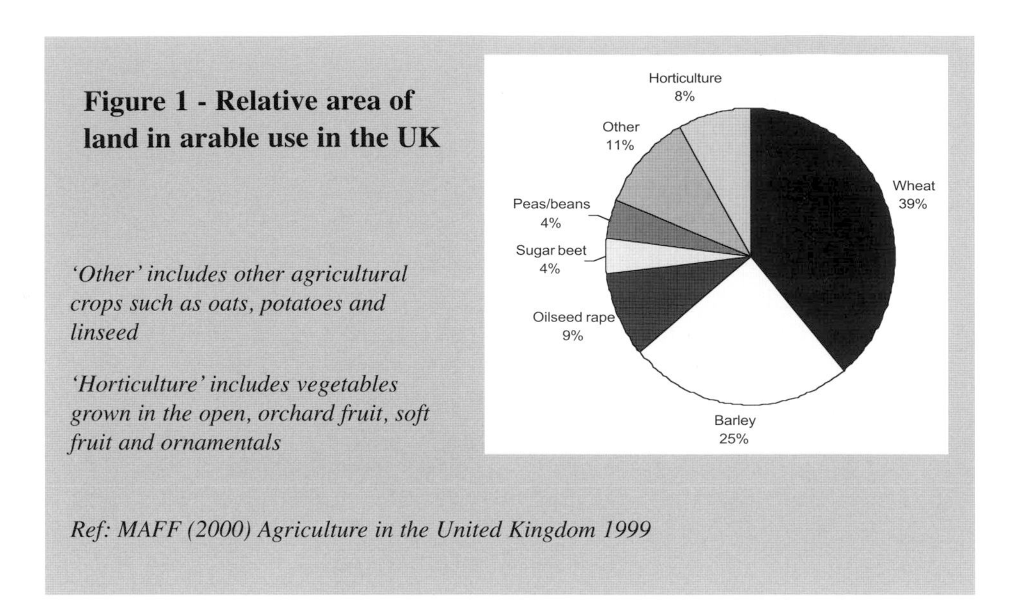

Figure 1 - Relative area of land in arable use in the UK

'Other' includes other agricultural crops such as oats, potatoes and linseed

'Horticulture' includes vegetables grown in the open, orchard fruit, soft fruit and ornamentals

Ref: MAFF (2000) Agriculture in the United Kingdom 1999

organisms, but hexaploid (i.e. has six sets of chromosomes), and domesticated maize has no seed dispersal mechanisms (Robinson and Treherne, 1990) and would die out without man's intervention.

Animal husbandry has resulted in dozens of breeds of cattle, pigs and sheep in the UK alone. In the case of cattle and sheep, it seems that the original wild ancestor has become extinct. Ironically, many of these breeds are now themselves close to dying out and specific conservation programmes are being introduced to prolong the line.

Traditionally, these breeding programmes have involved many years (sometimes hundreds of years) of crossing different varieties or existing breeds and selecting the offspring with particular preferred characteristics (Roeder, 1991). Significant strides were made during the agricultural revolution in Britain through breeders such as Robert Bakewell and Thomas Coke of Holkham. For example, in 1710 the average weight of an adult ox was under 400lb, but by 1795 this had risen to around 800lb (Cunningham, 1991). With the discovery in 1900 of Mendel's work on the basis of heredity, and the ensuing improved understanding of genetics in the early twentieth century, significant strides were made in animal breeding using 'directional selection' - that is, successively selecting and breeding from animals or plants for specific characteristics. Farnsworth (1978) quotes two examples of how directed selection was used to modify specific traits.

- Over the period 1933 to 1965, the number of eggs laid per hen per year by White Leghorn chickens rose from 126 to 250

- Over 50 generations, a starting strain of corn (maize) with a protein content of 10.9% and an oil content of 4.7% was used to develop a 'high yielding' strain (producing 19.4% protein and 15.4% oil) and a 'low yielding' strain (with protein and oil contents of 4.9% and 1.0% respectively).

Launch of GM tomato paste

Genetic modification has been used to produce a 'slow softening' tomato. As the fruit is more robust there is less wastage during harvesting and post-harvest handling. Softening results partly from the breakdown of pectin which cements the cells together. This breakdown is carried out by the enzyme polygalacturonase. Genetic modification was used to reduce the levels of the enzyme in the fruit, thus slowing pectin degradation, giving a firmer fruit and pastes of better consistency. As the softening is reduced, the fruit can be left longer on the vine to ripen naturally and produce a fuller flavour. The paste made from these tomatoes went on sale in two UK supermarkets on 4th February 1996, each with one of the following statements on the packaging. "The benefits of using genetically modified tomatoes for this product are less waste and reduced energy in processing" and "This quality tomato puree has been produced from genetically modified tomatoes. This modification helps the farmer to harvest crop at the best time, which in turn leads to a more usable, ripe fruit. Less energy is used in processing these tomatoes compared to the non-modified types".

Reference:

Jones, J.L. (1996) Food biotechnology: current developments and the need for awareness. Nutrition and Food Science (6): 5-11.

Jones, J.L. (1999) The food, the fad and the technology. Biologist **46** (3): 144.

Table 3 - Geographical origin of some major crops

Species	Origin	Cultivated
Potato (*Solanum tuberosum*)	S. American Andes	Throughout temperate world
Maize (*Zea mays*)	Central America	Worldwide tropical, subtropical,
Soybean (*Glycine max*)	Asia	Warm temperate, esp. Americas
Wheat (*Triticum aestivum*)	Middle East	Temperate worldwide
Sugarcane (*Saccharum officinarum*)	Indo-China	Tropical, subtropical worldwide
Coffee (*Coffea arabica*)	Ethiopia	Latin America, India, Indonesia, East Africa
Olive (*Olea europea*)	Middle East	Europe, Africa, Americas, Australia
Cocoa (*Theobroma cacao*)	Upper Amazon	Central & South America Central Africa
Rice (*Oryza sativa*)	South Asia	Asia, Africa, Middle East, Americas
Banana (*Musa* spp.)	SE Asia/Western Pacific	Tropical, Subtropical
Oilseed rape (*Brassica napus*)	Europe	Canada, Europe, China, India
Groundnut (*Arachis hypogea*)	Andes	Semi-arid tropics, subtropics and warm temperate
Barley (*Hordeum vulgare*)	South Asia	Temperate world

References:

Anon (1990) Exploited plants: collected papers from Biologist. Institute of Biology, London.

Duddington, C.L. (1969) Useful plants. McGraw Hill.

Heywood, U.H. and Chant, S.R. (1982) Popular Encyclopaedia of Plants. Cambridge University Press.

Moore, D.M. (Ed) (1982) Green planet: the story of plant life on earth. pp. 238-248. Domestication of plants. Cambridge University Press. ISBN 0 521 24610 5.

Nowadays, genetic manipulation techniques offer the potential for the process to be significantly speeded up and to be expanded so that traits from different species can be transferred between unrelated organisms. For example, a bacterial gene can be introduced into a plant to confer pest resistance. The environmental, ethical and safety aspects of this technique have aroused much public debate, and there is no doubt that these aspects will have to be properly addressed before the technique is fully accepted by the European consumer.

Breeding has gone hand in hand with geographical redistribution (see Table 3). This has meant, for example, that the potato, a native of the Americas, is now a staple food grown throughout the temperate world, and soya beans, which derive from Asia, are now extensively cultivated in the Americas.

1.3 What is food made of?

The simple answer to this is chemicals - and lots of them. Food is derived from plants, animals and microbes which are, in essence, highly organised chemical systems. Many of the chemicals within food are essential for human life - others just happen to be associated with the material we actually want to eat. We choose to eat certain foods because the combination of chemicals they contain give it a pleasant taste and provide what we need to survive. The role of the different components in nutrition is discussed in detail in many textbooks, including Whitney *et al.* (1998).

The types of constituents of foods can be broadly summarised as follows:

- carbohydrates, fats and proteins to provide energy and the building blocks for growth, development, maintenance and survival
- vitamins and minerals which are needed in small amounts for the body to be able to function properly
- fibre (along with other constituents) to assist in digestion and other functions

- flavour compounds, which generally serve no purpose other than to make the food taste good

- other compounds, many of which have no function at all as far as the human body is concerned but were required by the plant or animal during its lifetime

Obviously, we have developed knowledge to eat those materials that best provide the chemicals we need, but all foods will naturally contain substances that we don't need, and in some cases some that we would rather not consume at all. Cooking and different processing techniques can significantly alter the composition, content and availability of the chemicals in foods, as well as the food's physical and sensory properties. The major nutritional ingredients of some typical but different foods are shown in Table 4.

Table 4 - Major nutritional ingredients of selected foods
(Typical values)

Food type	Water %	Prot %	Fat %	Carb %	Starch %	Sugar %	Fibre(*) %
Beef, rump steak, raw	66.7	18.9	13.5	0	0	0	0
White bread	37.3	8.4	1.9	49.3	46.7	2.6	3.8
Potatoes, new, raw	81.7	1.7	0.3	16.1	14.8	1.3	1.3
Cheddar cheese	36.0	25.5	34.4	0.1	0	0.1	0
Eggs, chicken, raw	75.1	12.5	10.8	Trace	0	Trace	0
Apples, eating, raw	84.5	0.4	0.1	11.8	Trace	11.8	2.0

(* by the Southgate method)

Reference:

Holland, B., Welch, A.A., Unwin, I.D., Buss, D.H., Paul, A.A. and Southgate, D.A.T. (1991) McCance and Widdowson's Composition of Foods. 5th Edition. Royal Society of Chemistry and MAFF.

Carbohydrates

Carbohydrates are the principal source of energy in the human diet. Complex carbohydrates such as starch (a polysaccharide and the main carbohydrate source in the diet) are built up of individual sugar molecules (monosaccharides). There are several types of monosaccharides, but for energy-generating purposes they are all converted to glucose in the body, which is then metabolised, eventually, to carbon dioxide and water. We also ingest much carbohydrate in the form of sucrose (table sugar), a disaccharide which is composed of one molecule of glucose and one of fructose. A fairly constant supply of carbohydrate in the diet is essential, as the brain has a requirement for glucose as an energy source and cannot function on fats as a sole energy source.

Fats

Fats or lipids are the second important source of energy in the diet. On a weight-for-weight basis they yield over twice as much energy as carbohydrates. There are many different types of lipid, and they are utilised in many different ways by the body - contributing to cell, tissue and organ structures. Although they have a poor reputation, they are an essential part of the diet. There are certain lipids that the body cannot make itself and for which it is totally reliant on a dietary source. For example, two fatty acids that the body cannot make are linoleic and alpha-linolenic acid: these are the 'parent' acids of the omega-6 and omega-3 groups respectively, which play specific roles in body metabolism. Others lipids, such as cholesterol, can be made by the body, which also monitors and controls its level. In fact, dietary cholesterol has very little effect on the level of cholesterol in the blood. Fats are also needed to supply the essential fat-soluble vitamins, without which normal metabolism would not take place.

Proteins

Proteins can be used as an energy source for the body, if required. However, their primary function is in the formation of body structures, such as muscle. Proteins are made up of amino acids, of which there are 22. The body requires these in various quantities and cannot easily store them. A shortage of any one particular amino acid will mean that building work on many proteins and an entire particular body structure is halted. Therefore it is important to eat a mixture of proteins that will provide enough of each of the 22 individual amino acids.

Enzymes, which enable all living processes to take place under the kinds of conditions found in living cells, are also proteins. Dietary proteins are therefore ultimately required for their production also.

Fibre

Fibre is essential for the maintenance of correct function in the gut. Amongst other roles it is used by the bacteria in the gut, which contribute to the digestion of food before the individual components are absorbed. There is also evidence that certain types of fibre can affect the way that fat is transported in the bloodstream.

Minerals

The body needs a myriad of elemental minerals for correct function. Among these are potassium, sodium, calcium, zinc, phosphorus, magnesium, iron, manganese, selenium, and iodine. Zinc, for example, has structural, regulatory and catalytic roles in a number of enzymes, and also has a structural role in non-enzymic proteins. A deficiency of zinc can result in growth retardation and defects in the skin, intestinal mucosa and immune system (Department of Health, 1991). Some minerals are required in greater quantities than others and this requirement can vary considerably from one part of the population to another. For example, adolescents and breast-feeding women require more dietary calcium than other adults. The levels of minerals required per day are, generally, in the milligram range, and the body has several mechanisms for maintaining the correct balance of each individual mineral.

Vitamins

Like minerals, vitamins are required in small amounts. There are about 12 major vitamins which the body uses to facilitate a healthy metabolism. Deficiency in one or more of these vitamins can lead to serious illnesses. Different foods can be good sources of one or more vitamins, but it is important to take into account the fact that some vitamins (specifically the B group vitamins and vitamin C) are not very stable to processing or cooking, though severe vitamin deficiency is uncommon in the developed world.

Flavour and taste compounds

In addition to the major components of foods, described above, the food will also contain a range of flavour and taste compounds. These may have no role in nutrition at all, but they do actually help to persuade us to eat the food that we need to eat, and are therefore quite important. Processing and storage of food will actually change the nature and level of flavour compounds in a food - raw fruit and vegetables will have a different flavour from the canned equivalent, and cured meat will be different from the uncured product. Flavour compounds are numerous and complex; for example raw fruits often contain two or three hundred compounds that contribute to a greater or lesser degree to their overall flavour.

Other things that just happen to be there

These are exactly this. All food is of animal, plant or microbial origin. As such, the original living organism manufactured and used many chemicals that are not required by the organism that consumes it. Most of these are completely innocuous, having no role in or effect on human nutrition. Some have specific properties that maintain the structure and composition of the food (e.g. pectin in fruit). Some are actually beneficial to human health (such as the antioxidants found in soybeans, tea, red wine, spices, fruits, onions and olives - Langseth, 1995). Others that may cause health problems, we have learned to deal with. A common example of this is the lectins in kidney beans; these have to be rendered harmless by vigorously boiling the beans before eating them. A more extreme example is puffer fish, a delicacy in

Japan, which must be prepared in a precise fashion as it contains a lethal toxin. Others we just live with - like the compounds in onions that make us cry.

Additives

In addition to the chemicals that naturally make up the raw materials that we eat, other chemicals may be added during the processing or preservation of food - for a variety of reasons.

Although classed separately from other additives in legislation, flavours are the biggest and most widespread group of chemicals added to food. These may be of biological origin (i.e. extracted from an animal or plant - often termed 'natural') or synthetically produced via industrial processes. The synthetically produced chemical may, in fact, be exactly the same as one that occurs naturally - i.e. 'nature-identical'. It is quite common for industrially produced flavour mixtures to have up to 50 individual chemicals in different proportions to provide the exact flavour required. This gives an interesting comparison with the hundreds of compounds that are often found in the essential oils of fruits.

As well as flavours, there are many other classes of additives that are routinely used in processed food production (see Table 5). As with flavours, these may be synthetic, 'natural' or nature identical. These chemicals fulfil a variety of roles in foods (Table 5) and their use is very closely controlled by legislation. All additives permitted for use in the European Union are evaluated from a toxicological point of view; they are only permitted in foods where there is a proven technical need, and then only at levels that would not be expected to give rise to health concerns. Far from being 'bad', the E number system is proof that the compound has undergone these rigorous evaluations. In the USA, a similar and equally stringent approval system is operated.

Finally, other additives appear in food because they were used in the processing of the food. These processing aids have no function in the final product, and are merely 'left over' after processing because it is not feasible to remove them (e.g. lubricant on a sausage casing that facilitated the extrusion of the sausage).

In summary, food is largely derived from other living organisms - mostly organisms bred for the specific purpose of providing food. The food harvested or derived from the plant or animal is entirely chemical in its composition. Most food chemicals are either good for us or are of no nutritional consequence, but some - a small minority - are potentially harmful and represent a hazard that has to be controlled (see Chapter 4).

Table 5 - Major categories of food additives

Acid	Foaming agent
Acidity regulator	Gelling agent
Anti-caking agent	Glazing agent
Anti-foaming agent	Humectant
Antioxidant	Packaging gas
Bulking agent	Preservative
Colour	Raising agent
Emulsifier	Sequestrant
Emulsifying salt	Stabiliser
Firming agent	Sweetener
Flavour enhancer	Thickener

1.4 Innovation and product development

Consumers in affluent countries can be quite demanding. Food not only has to be safe, it has to be acceptable in terms of overall flavour and texture (i.e. sensory attributes), and also in such matters as nutritional content and with regard to ethical issues such as vegetarianism, and organic methods of production and other environmental matters. These demands are not static and so a continual process of new product development and new product placement is required if a manufacturer is going to maintain market share. Different groups of consumers demand different types of food and so provision of variety, allowing individual choice, has become a key feature of the UK food industry.

Consequently, the food industry has to be highly innovative, and in the UK around 5-7,000 new products are launched each year (see 1.4.1). This constant innovation in product development is the lifeblood of the industry - new products are needed to replace those whose sales are declining, to extend market share in growing sectors and to create new niches which meet ever-changing consumer needs. Some products are specifically introduced for limited periods to meet seasonal needs (e.g Christmas or barbecue products). Many products survive for only a short time - they are unable to compete enough to build a market share. And a few become success stories that get widely copied - just consider the growth in the market for sandwiches over the last few years.

1.4.1 Product trends

The food industry has seen major changes over the last ten or so years: the emergence of the major retailers and their own brands which has gone hand-in-hand with a consumer drive for the 'one-stop-shop'; the growth in convenience foods (e.g. microwaveable ready meals); the use of specific ethical or nutritional claims (e.g. vegetarian, low fat); and a huge proliferation in choice (the typical supermarket now stocks up to 20,000 different products). Meeting consumer desires and new commercial pressures means that food companies not only have to invest in the technological skills of product formulation and process development, but also in consumer research (e.g. understanding consumer attitudes) and marketplace intelligence (i.e. identifying which products appear and when).

New product trends in the UK market

The number of new food and drinks products (excluding confectionery and alcoholic drinks) launched each year more than doubled between 1991 (3233 new products launched) and 1999 (7318 new products launched), as shown below.

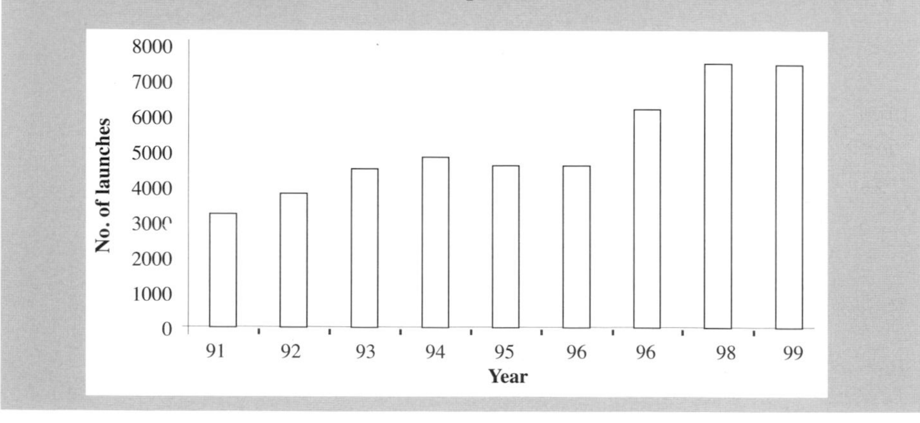

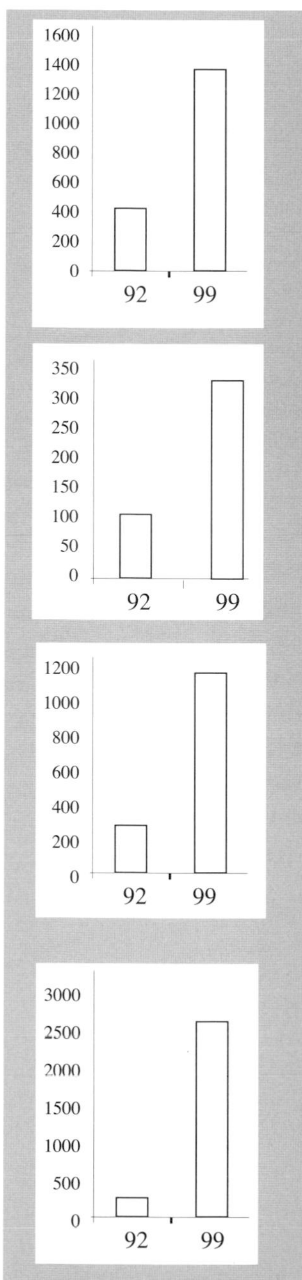

The annual number of new products suitable for microwaving and carrying microwave cooking or reheating instructions trebled between 1992 and 1999.

The annual number of new 'Sandwich Products' in 1999 was also around three times that in 1992.

Health-related claims such as 'low-fat', 'reduced sugar' and 'low salt' have become very prevalent. The annual number of new products carrying such claims increased four fold between 1992 and 1999.

The growth of vegetarianism as a main-stream consumer demand has been even more dramatic, and has been supplemented by a large number of non-vegetarians looking for meat-free products to vary their diet. The annual number of new products claimed as being suitable for vegetarians increased 10-fold between 1992 and 1999.

Reference:

Harrison, M., Llewellyn-Davies, D. and Everitt, M. (2000) New Products 1999 - 12 Month Review. CCFRA.

1.4.2. Product development

The time and money invested by major food producers and retailers in product development is significant and therefore enough products have to be successful in order to make the investment worthwhile. Shaw (1996) describes in detail the many steps involved in bringing a new product to the marketplace (Figure 2).

Figure 2 - The stages involved in product development

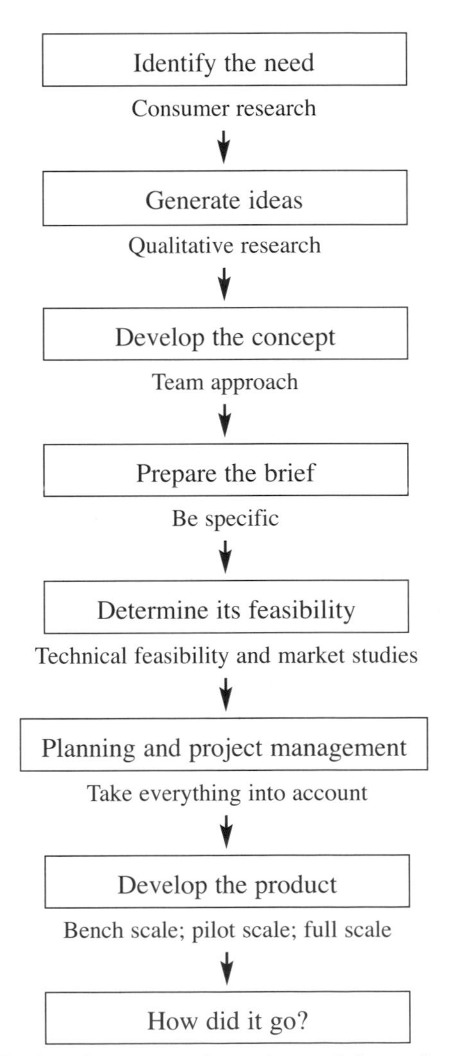

Product development - some considerations

Being aware of developments in the marketplace and knowing what consumers want are keys to success in product development. To obtain the most informative and useful data from consumers, a variety of traditional and novel market research procedures can be used: consumer group discussions, in-depth interviews, home placement trials, street interviews, hall tests and mail or telephone surveys. These are used to establish consumer expectations and needs, preferences and product acceptability by focussing on key attributes of the product, or key features of product concepts. This information can then be related to product profiles determined by trained panels to highlight the sensory characteristics of most importance to product acceptance, and hence help direct the product development process.

Understanding consumers, and what influences them, also needs information from the marketplace. It is important to identify trends in new products, to look for innovation, and to identify the social and economic drivers which will influence future consumption and buying patterns. The creation of a single European market for consumer goods has increased the importance of consumer segmentation in analysis, and in particular the need to understand cross-cultural consumer preferences.

Alongside this consumer research is the need to develop all technical aspects of a new product. As well as recipe formulation, pilot trials, product optimisation and manufacture of samples for market research, this also requires input from experts in the fields of legislation, packaging, labelling, and raw material supply, as well as specialist knowledge from microbiologists and technologists. It is essential that all of the technical aspects impacting on product manufacture are considered in the development stages. It is this integrated approach which ensures that the wider implications of all the factors which affect the product development process are fully considered. By considering all of the issues, companies have the best chance of getting it right first time: that is, developing a product that is technically feasible, that fulfils customer expectations and needs and is, most importantly, safe to consume.

1.5 The food production chain

The food production chain has become increasingly complex, as a consequence of the proliferation of choice for consumers, the development of markets for convenient and pre-prepared foods, the rapid movement of goods by modern transportation, faster communication facilitating global trade, and increasingly competitive markets (see Figure 3).

There is a net flow of material along the food production chain, from the primary producers (i.e. farmers and growers) through manufacturers and processors to retailers and food service outlets (caterers). A retailer will typically sell 'branded' and 'own-label' products side-by-side. A single formulated product, such as a ready meal assembled in the UK for example, might contain ingredients sourced from mainland Europe, Asia and the Americas. A single ingredient, such as soya bean lecithin, might pass through several hands between the farmer and food manufacturer. The farmer will harvest the beans which will be transported, warehoused, mixed with other beans, bought by brokers, sold on to suppliers, and then be processed and fractionated. These different fractions (e.g. oil, flakes) may in turn be sold separately before further fractionation by ingredients processors and suppliers.

While some ingredients are sourced in this way from the 'open market', others may be produced under contract - where a farmer and/or processor is contracted to a manufacturer to produce major ingredients (e.g. potatoes for snack product manufacture) to tightly defined and controlled specifications.

The sophistication of the food production chain has grown hand-in-hand with the emergence of quality and safety assurance systems (see Chapter 6). These will include, for example:

- specifications defining the quality of raw materials and end-products, which form the basis of trading agreements
- product and ingredient identification and traceability systems
- standard procedures to ensure products are manufactured in a consistent way to assure they deliver end-products of the quality required
- HACCP systems to assure product safety

These procedures reflect the move towards preventative approaches and have become an integral part of the way in which the many partners involved in the food product chain work together to develop and supply products that meet the needs and expectations of the consumer.

Figure 3 - The food production chain

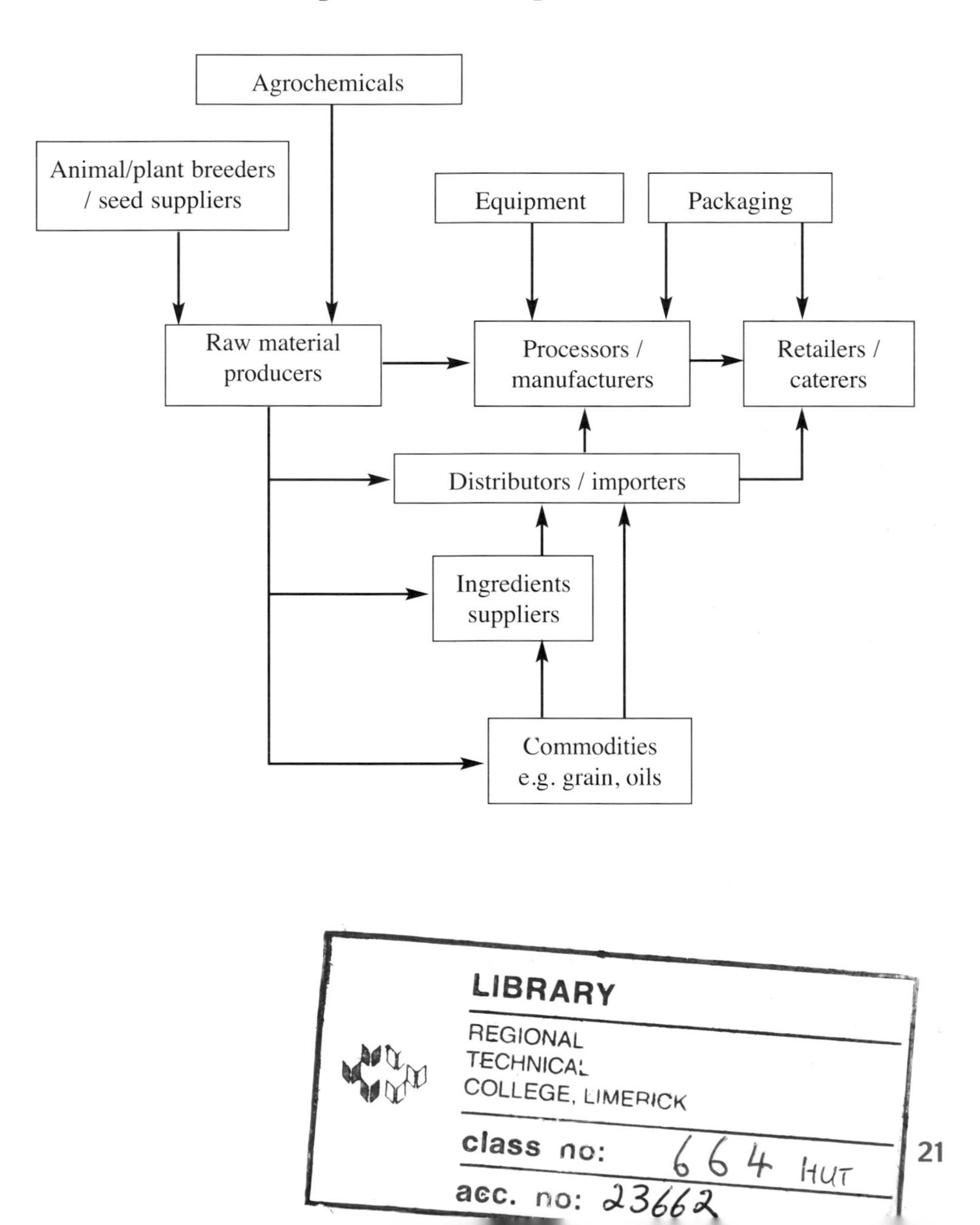

2. FOOD PRESERVATION AND PROCESSING TECHNOLOGIES

Processed foods have been produced and traded for millennia, although it is probably true to say that the vast majority of processed foods now traded use technologies developed over the last 100 years or so. Food is processed so as to alter its nature or sensory properties and/or to prolong the length of time it remains edible (its shelf-life) - that is, to make it palatable, to make it last, and to make it safe.

This chapter looks at the some of the principles underlying the major food processing operations and preservation technologies. The division between the two is not always clear or easy to define, and in industry the two concepts are often inextricably linked. Here we distinguish preservation technologies as concepts for prolonging the shelf-life of a food, and processing techniques as one or a series of unit operations for altering the physical and/or chemical nature of the food. Some techniques - such as canning and fermentation - combine the two concepts.

2.1 Food preservation

In most cases, it is the growth of either spoilage or disease-causing micro-organisms that limits the length of time for which a food can be kept, and most preservation techniques are primarily based on reducing or preventing this growth. However, there are other factors which limit shelf-life, such as the action of naturally occurring enzymes within the food, natural chemical reactions between the constituents of the food and physical processes such as drying, and these also have to be taken into consideration.

2.1.1 Heat processing

Micro-organisms and enzymes are the major causes of undesirable changes in foodstuffs; both are susceptible to heat, and different heating regimes can be used to reduce or destroy microbial and enzymic activity. The degree of heat processing required to produce a product of acceptable stability will depend on the nature of the food, its associated micro-organisms and enzymes, the conditions under which the heated food is stored, and any other preservation technique used.

Canning

The principal concept of food canning is to heat a product in an hermetically sealed container so that it is commercially sterile at ambient temperatures; in other words, so that no microbial growth can occur in the food under normal storage conditions at ambient temperature until the package is opened. Once the package is opened the preservation effects of canning will be lost, the food will need to be regarded as perishable, and its 'shelf-life' will depend on the nature of the food itself.

The most heat-resistant pathogen that has to be destroyed by the canning process is *Clostridium botulinum*. This can form heat-resistant spores under adverse conditions, which may germinate in the absence of oxygen and produce a highly potent toxin. As the canning operation results in anaerobic conditions (i.e. no oxygen) in the food, all canning processes take into account the need to destroy these spores. (See Section 4.1 for more details on this organism and its importance to the food industry).

In practical terms, the process applied must reduce the chances of a single spore surviving in a can to one in one million million (i.e. 1 in 10^{12}). This is called a 'botulinum cook', and should be such that the slowest heating point of the food in the container or the 'cold point' should receive the standard process of 3 minutes at 121.1°C or its equivalent. However, it is important to recognise that many factors inherent in the food may either allow a reduced level of processing to be applied, or necessitate an increased degree of processing.

In the traditional canning process, the product is filled into aluminium, tin plated mild steel or tin-free steel cans, lids are attached to form an hermetic seal, and the cans are heated under pressure in a retort. Heating is usually achieved by superheated water or steam, and care must be taken to ensure that the heat penetrates to the centre of the can and is distributed evenly, so that no 'pockets' of product are left underprocessed. At the same time, it is desirable not to overcook the product, as this will result in reduced product quality. Therefore, much research has been undertaken to find the best ways of assuring adequate heat penetration and distribution, and of measuring these.

After heating, the product needs to be cooled under controlled pressure and temperature; it is vital that no post-process contamination occurs through the can seals. Therefore, the integrity of the seal is vital and there are strict regimes for post-process can handling. The water used to cool the cans must be of high quality microbiologically (i.e. potable), and the cans must not be handled while still hot and wet as this could potentially result in contamination, with the water acting as a conduit for any micro-organisms present. The cause of an infamous typhoid food poisoning outbreak in 1964 in Aberdeen, linked with corned beef, was traced back to post-process contamination. The cans had been cooled in the open air with water from the local river, downstream from a military fort where a typhoid outbreak had occurred! There are anecdotal reports of people still not buying corned beef because of this incident.

It is important that the cans are not damaged or deformed, as this could result in either perforation of the body of the can or damage to the lid seal. Either, when combined with a wet can surface, could allow the entry of micro-organisms into the can, thus negating the effects of the sterilisation process.

Although cans are the traditional form of packaging for this type of process, glass vessels and flexible pouches can also be filled and processed in a similar way, and are often used.

Aseptic (continuous) processing

As in canning, the aim is to produce a commercially sterile product through heat processing. The major difference here is that the package and product are sterilised separately and then the package is filled aseptically (i.e. under conditions which exclude micro-organisms). Unlike canning, which is a batch process, product is sterilised in a continuous process in which it travels through a heating pipe and holding tube before entering a cooling leg and being transferred to the package. This technique is particularly suitable for fluid foods such as soups and for fruit juices and similar products. One potential advantage is that of enhanced product quality, as the potential problem of overcooking can be significantly reduced. There is also the scope for the use of a range of package types, such as cartons, plastic pouches and glass jars (see Chapter 3).

Pasteurisation

This is a heating regime (generally below 105°C) that aims to eliminate all pathogens (apart from spore forming organisms), as well as some spoilage organisms, from a foodstuff. The actual degree of heat process required for an effective pasteurisation will vary depending on the nature of the food and the nature and number of micro-organisms present. Milk is the most widely consumed pasteurised food in the UK and the process was introduced in this country in the 1930s, when a treatment of 63°C for 30 minutes was used. Modern milk pasteurisation uses an equivalent process of 72°C for 15 seconds (Singleton and Sainsbury, 1978; Singleton, 1997). Pasteurisation is now used extensively in the production of many different types of food, notably ready meals. Food may be pasteurised in a sealed container (analogous to a canned food), or in a continuous process analogous to an aseptic process. It is important to note that pasteurised foods are not sterile and will usually rely on other preservative mechanisms to ensure their extended stability for the desired length of time. Chilled temperatures are most commonly used, but some products have a high enough salt or sugar content to render them stable at room temperature (e.g. pasteurised 'canned' ham). Whereas canned, sterilised foods as described above have shelf-lives measured in years, pasteurised chilled foods rarely have shelf lives of more than 2-3 months, and in many cases significantly less than this.

2.1.2 Freezing

Freezing of food does not render it sterile. Although the freezing process can reduce the levels of some susceptible micro-organisms, this is not significant in the overall microbial quality of the food. However, at the freezing temperatures used, all microbial activity is suspended and the length of time for which the product can be kept is dependent on other factors. It is important to note, however, that once a frozen food is defrosted, any micro-organisms present will be able to grow and multiply.

Even at freezing temperatures, some enzymic activity may remain, which could alter the sensory properties of the food. The potential problems with enzyme activity will depend on the food in question, but peas, for example, will quickly develop off-flavours if frozen without pre-treatment. Therefore, vegetable products are usually blanched before freezing to ensure that enzymes are inactivated. A typical blanching process involves heating the product to 90-95°C for a few minutes before rapid freezing. The rate of freezing is important. Rapid freezing in blast freezers is desirable to prevent the formation of large ice crystals which will tend to adversely affect the texture of the product.

Apart from enzymic activity, there are many other chemical and physical changes which may limit the shelf-life of frozen food; examples include fat oxidation and surface drying, both of which may occur over a period of months, depending on the product.

2.1.3 Chilling

Keeping products at a low temperature retards both microbiological and chemical deterioration of the food (see Table 6). In most processed chilled foods, it is microbial growth that limits the shelf-life; even the slow growth rates that occur under chilled conditions will eventually result in microbial levels that can affect the product or present a potential hazard. This microbial growth can result in spoilage of the product (it may go putrid or cloudy or show the effects of fermentation), but in many cases pathogen growth may have occurred with no noticeable signs of change in the food. It is important to note that food showing microbial spoilage is not necessarily poisonous. Conversely, just because a food has not spoilt does not mean that it is safe, and it is important to pay attention to the 'use-by' date on chilled foods.

To reduce microbial effects to a minimum, chilled prepared foods are usually given a pasteurisation heat process, sufficient to eliminate a variety of pathogens such as *Salmonella*, *Listeria* and *Escherichia coli* O157. A process equivalent to 70°C for 2 minutes is generally considered to be sufficient, but the exact process given will depend on the nature of the food. After processing it is essential to prevent recontamination of the product, which means producing the food in a closely controlled 'high care' area (see Ridgwell, 2000; Stringer and Dennis, 2000).

Chilling is also used to prolong the shelf-life of fresh fruit and vegetables. Here, low temperatures not only retard the growth of naturally occurring micro-organisms (which might rot the product), but also slow down biochemical processes that continue after the product has been harvested. However, each individual fruit and vegetable has its own ideal storage temperature, and some are susceptible to chill injury (such as bananas - see Chapter 5.1)

Table 6 - Minimum growth temperatures for selected pathogens

Organism	Minimum Growth Temperature (°C)
Bacillus cereus	4.0
Clostridium botulinum (psychrotrophic)	3.3
Escherichia coli O157	7.0
Listeria monocytogenes	-0.4
Salmonella species	4.0
Staphylococcus aureus	6.7
Vibrio parahaemolyticus	5.0
Yersinia enterocolitica	-1.0

Reference:

Betts, G.D. (1996) A code of practice for the manufacture of vacuum and modified atmosphere packaged chilled foods with particular regard to the risks of botulism. Guideline No. 11. CCFRA.

2.1.4 Drying and water activity

Micro-organisms need water to grow. Reducing the amount of water in a product that is available for the micro-organism to use is one way of slowing or preventing growth. Thus, dried herbs and spices may be microbiologically stable for many years. Many staple foods are available in dried forms (e.g. breakfast cereals, pasta and rice) and these will also be stable for a long period of time. The shelf-life of breakfast cereals is usually limited by texture changes, with the product losing its crispness and becoming soft or 'cardboardy'.

Some products such as jam may contain fairly high levels of water, but much of this is 'tied up' by the sugar present in the jam (i.e. the product has a low water activity) and is not available for the micro-organism to use. As a result, traditional jams can be kept for many months without spoiling. Conversely, many low-sugar jams have to be refrigerated after opening, as the sugar levels are not sufficient to prevent microbial growth. This is a good example of how the consumer-driven desire for products with altered characteristics (i.e. jam with less sugar and no 'artificial' preservative) has resulted in a product which has lost one of its major, original characteristics - long-term stability at room temperature.

Other products may have a relatively low water content, but that which is present is freely available for microbial growth (e.g. cheese).

2.1.5 Chemical preservation

The addition of specific chemicals to foods to inhibit microbial growth and chemical transformations is a major method of preserving food. Herbs and spices have been used traditionally, albeit originally unwittingly, for this purpose. Antimicrobial additives ('preservatives') probably receive most attention; there are relatively few of these permitted for use in the UK and EU and in many cases there are specific limits on how much can be used and in which products. The use of some preservatives is limited to just a few types of food (e.g. nitrate and nitrite salts to specific meat, cheese and fish products). The two preservatives used most widely are sorbic acid and its salts and sulphur dioxide and its derivatives. There is currently a major consumer-led move to reduce the range of foods containing preservatives and the levels of preservatives actually used. This poses a significant problem for the food industry, as a reduction in preservative level either necessitates another preservation technique to be used (e.g. heating or freezing) or a significant reduction in shelf-life. Both of these alternatives may result in a product of actual or perceived poorer quality.

In addition to preservatives, antioxidants are widely used to prevent chemical deterioration of foods; this includes the rancidity caused by oxidation of fats, and the browning of cut vegetables caused by the formation of polymers after the action of the enzyme polyphenol oxidase.

Some forms of chemical preservation are well-established, traditional techniques, as outlined below.

Curing

Strictly speaking, curing actually means saving or preserving, and processes include sun drying, smoking and dry salting. However, curing is now generally thought of as the traditional process that relies on the combination of salt, nitrate and for nitrite to effect chemical preservation of the food, usually meat, but also to a lesser extent fish and cheese.

In the curing of meat, the salt has preservative and flavour effects, while nitrite also has preservative effects and contributes to the characteristic colour of these products. Typical cured products are bacon, ham and gammon, and there are a number of variations on the curing technique. Ranken *et al.* (1997) describe the traditional Wiltshire curing of bacon, which involves injection of brine into the pig carcass, and immersion in a curing brine which contains 24-25% salt, 0.5% nitrate and 0.1% nitrite. The curing brine is used from one batch to another, being ‘topped up’ between batches, and is a characteristic deep red colour due to the high concentration of protein that accumulates.

Pickling

This commonly refers to the preservation of foods in organic acid or vinegar, although the term can be applied to salt preservation. Most food poisoning bacteria stop growing at the acidity levels (pH 4) attained during the pickling process, although yeasts and moulds require a much higher degree of acidity (pH 1.5-2.3) to prevent growth.

A number of vegetables are pickled in vinegar in the UK, such as beetroot, gherkins and cucumbers, onions and cabbage, as well as walnuts and eggs. Mixed pickles are also available. In some products, the raw or cooked material is simply immersed in vinegar to effect preservation, but in others, additional processes such as pasteurisation are required to produce a palatable and safe end-product, e.g. pickled beetroot is peeled and usually pasteurised after cooking.

The march of technology

It is around 200 years since Nicholas Appert invented canning. After James Lind had demonstrated a link between scurvy and diet, the British Admiralty adopted canned foods with gusto. In 1851, however, a large quantity of naval stock was found to be putrid - a condition which arose as a result of inadequate heat processing of the batches concerned. This led to alarm about the canning process in general. Newspapers published a flood of correspondence - explanations were offered, experiences recounted and public enquiries demanded. By the 1920s, when canning of food for general sale really became established in the UK, reservations about this 'new technology' rumbled on. Writing in Good Housekeeping in 1923, William Savage MD attempted to lay out the facts about canned foods in his article 'The Truth about Canned Foods' which opens:

'There exists a widespread prejudice against preserved foods generally, and against tinned or canned foods in particular, which is fanned by much ill-informed newspaper and other criticism ......' and goes on to say: *'It is necessary to realise that we have to discriminate between the dangers to health which are peculiar to this particular method of preserving food and dangers which are common to all foods, preserved or not.'*

Despite Savage's reassurances, which have interesting parallels with the debate in the 1970s on irradiation and in the late 90s about GM foods, the debate evidently continued. In 1929, the group of notable medics and academics making up the Food Committee of the New Health Society were prompted to publish, also in Good Housekeeping, a summary of their report into canned foods which asserted that:

'Experience proves that modern meticulous methods of harvesting, selection, preparation and canning guarantee absolute safety.'

Fortunately, canned foods were not consigned to history, and new technologies continue to emerge. For example, modified atmosphere packaging (MAP) is now widely used for certain foods (see 2.1.7). This is a system where the atmosphere surrounding the food is adjusted by flushing with a controlled composition of gases, to inhibit deleterious reactions. 'Active packages' take this a stage further and can contribute to food preservation by providing more than just a barrier to contamination. For example, some carry materials which scavenge ethylene (a natural plant growth hormone) from the atmosphere around the product, to reduce deterioration through over-ripening. High pressure processing could offer better retention of colour, flavour, texture and nutritional properties of some raw materials because it does not involve heat, and genetic modification offers routes to developing foods for those with special dietary needs (e.g. gluten-free wheat for coeliacs, who have a gluten intolerance).

References:

Braithwaite, B. and Walsh, N. (1990) Food glorious food: eating and drinking with Good Housekeeping 1922-42. Ebury Press. ISBN 0 85223 818 5.

Jones, J.L. (1999). The food, the fad and the technology. Biologist **46**(3): 144.

Thorne, S. (1986). The history of food preservation. Parthenon Publishing.

Smoking

This is another traditional process which relies on chemicals to effect preservation of the product. Meat smoking derives from the practice of hanging meats in the chimney or fireplace to dry out. This had a variety of effects: the meat was partially dried, which itself assisted with preservation, but chemicals in the smoke also had direct preservative and antioxidant effects, as well as imparting a characteristic flavour on the product. In modern smoking techniques, the degree of drying and of smoke deposition and cooking are usually controlled separately. In cold smoking, there is no cooking step, whilst in hot smoking, a cooking element is included.

Smoked salmon is an example of a product in which brining is used in combination with smoking to give an end-product with an extended shelf-life.

2.1.6 Fermenting

Many foods are fermented. In fermented foods, preferred micro-organisms are allowed or encouraged to grow in order to produce a palatable, safe and relatively stable end-product. Those micro-organisms present prevent or retard the growth of other undesirable spoilage or pathogenic organisms, and may also inhibit other undesirable chemical changes. Some of the main types of industrial fermentation are: bacterial fermentation of carbohydrates, as occurs in yoghurt manufacture; bacterial fermentation of ethanol to acetic acid (as in vinegar production); yeast fermentation of carbohydrates to ethanol (as in beers, wines and spirits); and yeast fermentation of carbohydrates to carbon dioxide (as in bread doughs). Many fermented milk products involve lactose (milk sugar) fermentation to lactic acid. In many cases the fermentation results in the production of chemicals (e.g. acetic acid or ethanol) which act as chemical preservatives.

2.1.7 Modified atmosphere packaging (MAP)

This is a technique that is being increasingly used to prolong the shelf-life of fresh foods such as meat, fish, fruit and vegetables, as well as various bakery products, snack foods and other dried foods. The basic idea is to replace the air in a package with a gas composition that will retard the deterioration of the food. Clearly, the properties of the packaging material and film will be important in maintaining the gas composition. In most cases, it will be microbial growth that is inhibited by the modified atmosphere, but in some situations, the onset of rancidity and other chemical changes can be delayed. The exact composition of the gas used will depend entirely on the type of food being packaged and the biological process being controlled (see Air Products, 1995; Day, 1992a). MAP is generally used in combination with refrigeration to extend product shelf-life of fresh, perishable foods. Table 7 gives some typical gas mixtures used.

Table 7 - Typical gas mixtures use in MAP of retail products

Product	O_2	CO_2	N_2
Raw red meat	70%	30%	
Raw offal	80%	20%	
Raw, white fish and other seafood	30%	40%	30%
Raw poultry and game		30%	70%
Cooked, cured and processed meat products		30%	70%
Cooked, cured and processed fish and seafood products		30%	70%
Cooked, cured and processed poultry products		30%	70%
Ready meals and other cook-chill products		30%	70%
Fresh pasta products		50%	50%
Bakery products		50%	50%
Hard cheeses		100%	
Other dairy products			100%
Dried foods			100%
Liquid foods and drinks			100%

Reference:

Air Products (1995) The Freshline guide to modified atmosphere packaging.

There are several techniques related to MAP that are worthy of mention. Unprocessed fruit and vegetables continue to respire after being packed, consuming oxygen and producing carbon dioxide. Using packaging with specific permeability characteristics, the levels of these two gases can be controlled during the shelf-life of the product. Alternatively, 'active' packaging can be used in which chemical adsorbents are incorporated, e.g. to continuously remove oxygen or water from the package as they are produced during storage.

As an alternative to controlling or modifying the atmosphere, in vacuum packaging all of the gas in the package is removed. This can be a very effective way of retarding chemical changes such as rancidity development, but care needs to be taken to prevent the growth of the pathogen, *Clostridium botulinum*, which grows under anaerobic conditions (see 3.2).

2.1.8 Novel techniques

Food technologists are continually looking for new ways to produce food with enhanced flavour and nutritional characteristics. Traditional thermal processes do tend to reduce the vitamin content of food and can also affect its flavour. The development of processes which are as effective as traditional thermal systems in reducing or eliminating micro-organisms, but do not adversely affect the constituents of the food, are being actively developed. In addition to those mentioned below, ultrasound, pulsed light, microwaves, and electric field and magnetic field systems are all being actively investigated. In the UK, before any completely novel food, ingredient or process can be marketed, it has to be considered by the Advisory Committee on Novel Foods and Processes. The primary function of the ACNFP is to investigate the safety of the novel food or process and to advise government of their findings. The European Union has also formulated community-wide ‘Novel Foods’ legislation.

High pressure processing

This technique was originally thought of in the 1890s, but it was not until the late 1970s that Japanese food companies started to develop its potential. Pressures of several hundred times that of atmospheric pressure are used, in cycles of short bursts, to kill unwanted micro-organisms. Jams were the first commercial products to be produced in this way in Japan, and the process is now being investigated in Europe and the USA.

A Spanish company, Espuna, has recently started commercial production of pasteurised ham while, in France, a Pernod Ricard subsidiary is producing freshly squeezed orange and grapefruit juices.

In the USA and Mexico, Avomex launched guacomole for chilled distribution in 1996 and other companies have looked into commercialising other seafood, spreads and fruit juices.

Ohmic heating

This is a thermal process, but instead of applying external heat to a product to create a sterilisation effect, an electric current is applied directly to the food; the electrical resistance of the food to the current causes it to heat it up. The advantage is that much shorter heating times can be applied than would otherwise be possible, and so the product will maintain more of its nutritional and flavour characteristics. Frankfurter style sausages are particularly suited to this type of process.

Irradiation

This technique has seen much wider applications in the USA than in the UK, where 'public opinion' has effectively sidelined it. In addition to killing bacterial pathogens, such as *Salmonella* on poultry, it is especially effective at destroying the micro-organisms present on fresh fruit such as strawberries and thus markedly extending their shelf-life. It can also be used to prevent sprouting in potatoes. Its biggest advantage is that it has so little effect on the food itself that it is very difficult to tell if the food has been irradiated. However, in practice it is an expensive process to carry out and commercial applications are generally restricted to low volume/high value products such as herbs and spices. It also has some technical limitations, in that it is not suitable for products that are high in fat, as it can lead to the generation of off-flavours. The only commercial food products that are currently licensed for irradiation in the UK are dried herbs and spices, which are difficult to decontaminate by other techniques, without markedly reducing flavour.

In the UK there is a requirement to label food that has been irradiated or contains irradiated ingredients.

2.2 Food processing

In addition to preservation methods primarily designed to enhance the shelf-life of foodstuffs, there are a myriad of other processing techniques that are needed to turn raw materials into final products, and to formulate the food in the desired way. Some of these yield end products (such as cheese, bread and cakes) that bear little or no physical resemblance to the starting materials.

2.2.1 Raw material preparation

Whether the raw materials be animal or vegetable in origin, they need to be cleaned and treated in some way before they can pass 'down the line' to be further processed or passed directly to the consumer. The type of treatment needed will depend on the type and origin of the raw material. Meat and poultry will need to be slaughtered, cleaned, butchered and dressed and fish will need to be filleted and cleaned. Both of these groups of operations usually take place in specialist establishments (e.g. abattoirs), and detailed discussion on these is outside the scope of this document. Fruit and vegetable pretreatment, however, tends to be often carried out in the food manufacturing environment. The pretreatment of cereals can be divided into two steps: an initial clean-up phase, similar to that seen with fruit and vegetables, and the specialist milling stage, which is discussed later in this chapter.

Before fruit, vegetables and cereals can be further processed, they need to be cleaned and have any extraneous matter that might be present (such as soil, stones, insects, metal, glass and straw) removed from them. Vegetables from the soil are particularly likely to be contaminated. There are a number of decontamination techniques that can be used - from water flotation methods, air blowing and physical size separation to metal detectors, colour separating systems and sophisticated fault detection machines. These are discussed briefly in Chapter 4 and in more detail in Campbell (1995).

After being cleaned, the material may need to be sorted by size (e.g. with peas), or peeled (e.g. apples), chopped, diced (e.g. root vegetables) or trimmed (e.g. cauliflower and broccoli). The degree of this treatment will vary depending on the final destination of the raw material: fruit and vegetables for the fresh market may require little pretreatment after cleaning, whereas those that are going to be ingredients of a processed meal (e.g. fruit flan or savoury ready meal) may need more extensive treatment.

2.2.2 Generic processing technologies

To enumerate and describe all of the processing techniques used in the food industry is beyond the scope of this document. It is important to keep in mind that food manufacture involves the combination of many individual processes. Some of the major generic processes are briefly described below, and Table 8 gives simple definitions for a wider selection.

Cooking

This is probably the oldest food processing method known to mankind. It is, of course, a form of heat preservation, but its original primary role was to render food more palatable, and in many cases to produce an edible final product from an inedible starting material. Uncooked kidney beans, for example, are highly toxic, and some other vegetables, such as potatoes, are not particularly edible in the raw state. Its most significant effect in vegetables is to break down rigid cell structures and so make foods more tender. In high-protein foods such as eggs, meat and fish, cooking gelatinises the proteins, thus markedly altering the texture of the product. Most food consumed is cooked in some way, and many individual processing regimes contain cooking as an integral part.

Milling

If cooking is the oldest food processing technique known to mankind, then milling is probably the oldest process to take a raw material and prepare it for use as a staple food. In its simplest terms it is the grinding of wheat or other grains into a fine powdery flour (a size reduction process), but there are many factors that have to be considered if flour of the desired quality is to be produced. The first thing to consider is the quality of the raw material itself. There are many different varieties of wheat, which produce flours for different end products. From the miller's point of view, these have different milling characteristics: hard (bread) wheat endosperm is more crystalline and breaks into larger chunks, while that of soft wheats (for cakes and biscuits) is amorphous and crumbles into smaller particles (Posner and Hibbs, 1997). Storage facilities, handling systems and atmospheric conditions can significantly affect the quality and value of wheat, and these need to be carefully controlled, bearing in mind the likely starting quality of the wheat and end use of the flour.

In some cases it may be necessary to mix flours or grain of different characteristics. For example, high-quality breadmaking flour will usually be produced by milling a mixed grist of strong and weak wheats (Kent and Evers, 1994). Strong wheats generally have a high protein content and produce bread of a large loaf volume and good crumb texture, as the elasticity of the gluten allows expansion of the dough during the fermentation step of breadmaking (see also Section 2.2.3). Typically, more expensive strong wheat is blended with less expensive wheats which would otherwise yield a flour that would be too weak for breadmaking. The proportion of the wheats will vary from season to season (reflecting the seasonal variation in strength of the home grown wheat) and with the breadmaking process in use. In contrast, biscuit flour, for which an elastic dough is not desired, will use a weaker grist.

The grinding process itself can be classified into four parts: the break system, which separates the endosperm (from which the flour is derived) from the bran and the germ; the sizing system, which separates the small bran pieces attached to the large pieces of endosperm; the reduction system, which converts the endosperm to flour; and the tailings system, which separates the fibre from the endosperm recovered from the other three systems. After grinding, the material is sieved. This not only allows the flour to be separated, but also allows the other material to be classified for further processing.

Extrusion

Breakfast cereals, pasta, snack foods and texturised vegetable proteins are some of the major food groups that can be produced by extrusion techniques. The starting materials are primarily starch-based powders or granules, such as wheat, maize, rice and potato, or protein-based beans such as soya.

In the case of breakfast cereals or starch-based snack foods, cereal flours are mixed with water and other minor ingredients (such as sugar, salt and flavourings) to form a paste which can be squeezed through apertures to form desired shapes or pieces. These are then dried to the desired moisture content, and can subsequently be coated if required.

Emulsification

Emulsification is the dispersion of one liquid phase, in the form of fine globules, in another liquid phase with which it is incompletely miscible (Ranken *et al.*, 1997). In the food industry, the emulsions are generally either water-in-oil or oil-in-water systems. Among the many products which are emulsions in their final form are ice cream, margarine, butter and similar spreads (water-in-oil), and sauces, including mayonnaise and salad cream (oil-in-water). The ingredients of many other products (such as bread, cakes and chocolate) undergo an emulsification step at some point during manufacture.

The formation of an emulsion basically involves the use of mechanical energy to disperse fine globules of oil into a water matrix, or vice versa - the greater the energy input, generally the smaller the dispersed globules. In many situations, emulsifiers are used to increase the efficiency of emulsification. These are chemicals that are capable of dissolving in both water and oil, to some degree. They not only help in the actual formation of the emulsion, they also help to stabilise it. Some food ingredients themselves contain high levels of emulsifiers (both egg yolk and soya beans are high in lecithin, a phospholipid, and milk is high in casein, a protein), but synthetic or semi-synthetic food grade emulsifiers are also used by the industry.

Table 8 - Examples of processing operations

Aeration	the incorporation of air, usually into emulsified ingredients and usually requiring a stabiliser to maintain the desired level
Baking	cooking in an oven with dry heat
Blanching	a brief heat process designed to reduce or eliminate enzyme activity
Carbonation	the addition of carbon dioxide to drinks to make them fizzy
Clarification	of drinks; removing the cloud from fruit juices or alcoholic drinks by physical or chemical means
Coating	a variety of applications; e.g. putting breadcrumbs on fish fingers
Enrobing	e.g. the covering of confectionery with chocolate
Extraction	the removal of individual chemicals, or group of chemicals or ingredients from a raw material or intermediate
Filtration	the removal of solid particles from a liquid product
Glazing	e.g. the covering of fish products with ice or the covering of confectionery with a sugar solution
Granulation	e.g. of sugars and coffee
Homogenisation	the mixing of different components or phases of a food so that it is the same throughout, e.g. the water and fat components of milk
Hydrogenation	the conversion of unsaturated fatty acids (usually of plant origin) to saturated fatty acids
Moulding	the shaping of bread and confectionery products, amongst others
Panning	the application of many layers of coating to a centre tumbling in a revolving pan

Table 8 - Examples of processing operations (continued)

Puffing	e.g. of rice for breakfast cereals
Refining	reducing cocoa solid particle size in chocolate manufacture before mixing with fats
Rendering	the extraction of fats from animal tissues for lard and tallow production
Spray drying	a method of producing dry powdered products, e.g. milk powder
Ultrafiltration	filtration at the molecular level, whereby small molecules can pass through a molecular sieve, but larger ones are held back

Use of enzymes

Enzymes are used in a wide variety of food processing operations. In most cases, the enzymes are involved in hydrolysis (i.e. breakdown) of carbohydrates and proteins; three specific examples are in the tenderisation of raw meat, in the processing of fruit juices, and in the processing of milk products (Birch *et al.*, 1981). Also, as described in the next section, rennet is widely used in the production of cheese.

Tenderness is a highly desirable quality in meat, and is largely achieved by the natural enzymes remaining in the meat after slaughter; however, tenderisers can be added to the meat to augment this action. Papain, a plant enzyme which hydrolyses proteins, is the most commonly used, and interestingly, it exerts its actions mainly during the cooking of the meat, as it has its optimum activity at higher temperatures.

Enzymes are used in fruit juice processing for a variety of reasons: they help in the liquefaction of the fruit flesh and also in the extraction of colour and flavour components from the fruit; they are also used to clarify the fruit juice and to reduce its viscosity, making it easier to concentrate.

Lactase is an enzyme that splits the disaccharide lactose into its constituent sugars, galactose and glucose. It is used in the processing of milk for cheese and yoghurt manufacture, and also for the production of sweeteners (lactose itself is not particularly sweet). It can also be used to remove lactose from milk products and so make them suitable for those with lactose intolerance.

2.2.3 Specific food types

As stated above, some combinations of processes allow the production of foods that bear little resemblance to their starting materials. Some of these, such as flour and sugar confectionery, ice cream and soft and alcoholic drinks, have become part of our leisure and recreational activities.

With the current trend for reduced processing of foods, it is worth considering some of the more traditional foodstuffs and the extensive processing that goes into their production. Each involves the use of a combination of ingredients with unique functional properties. Minor modifications in the amount or type of each ingredient can result in significantly different products; this has resulted in the huge variety of breads, cheese, cakes, confectionery and alcoholic beverages that have been available for centuries. In some cases the final product as consumed takes several years to produce. A few examples are briefly described below.

Dairy products - cheese

There are thousands of different varieties of cheese available worldwide, and yet they all derive from the fermentation of milk. The basic production variables are the origin (e.g. cow, sheep, goat) and pretreatment (e.g. heat-treated or not) of the milk, the conditions under which the fermentation takes place (e.g. starter organisms, temperature, time, acidity and salt level) and subsequent ripening during storage of the cheese. Subtle combinations of these variables result in the many varieties of cheeses produced.

Milk is sometimes initially heat treated (a brief thermisation process at about 65°C for 15 sec) and is usually given a full pasteurisation (71.7°C for 15 sec) immediately before use. While this kills any pathogens present in the milk, it may also destroy some enzyme activity and cause other chemical changes which will affect the final quality of the cheese. Thus pasteurised and unpasteurised cheeses will have significantly different characteristics. Acidification is a key step in cheese manufacture; this happens through the production of lactic acid by micro-organisms in the milk. Originally, the indigenous microflora of the milk was relied upon for this, but nowadays starter cultures (e.g. *Lactococcus* and/or *Streptococcus* strains) are used, to allow easier control of the process. In some cheeses such as Mozzarella, acid is added to the milk (Fox, 1993). While acidification is occurring, the protein (casein) component of the milk is coagulated, usually by the addition of rennet. This results in the formation of a protein gel which also entraps milk fat, if present. This gel can be physically manipulated and its moisture content altered, to help give a cheese its specific characteristics; the addition of salt contributes to this.

Finally, most cheeses then undergo a period of ripening, sometimes of two years or more, during which a series of biochemical changes occur, due principally to enzymes originally in the milk, effects of the starter culture and its enzymes, and colonization by other micro-organisms. In some cheeses, the secondary micro-organisms provide its main characteristic; some are deliberately added, such as *Penicillium* mould to Camembert and Brie (Fox, 1993).

Dairy products - yogurt

Yogurt is fermented milk, and is believed to have originated at least 10,000 years ago (Tamine and Robinson, 1999), probably in the Middle East. Although there are around 400 generic names that are applied to fermented milk products around the world, they are all basically lactic fermentations, following essentially the same process steps. Firstly, the level of solids in the milk is raised to about 14-16%; the milk is then held at high temperature for 5-30 minutes, before being inoculated with a bacterial culture in which *Lactobacillus delbrueckii* subsp. *bulgaricus* and *Streptococcus thermophilus* predominate. The milk is then incubated to promote the formation of a smooth viscous coagulum along with the desired sensory properties. After cooling, fruit or flavourings can be added and the product can be pasteurised, if desired.

Cereals products - bread

Bread comes in many forms and can be made via a number of methods; in the UK, the production of the majority is based on the principle of mixing flour (usually wheat flour) and water with yeast and other ingredients to form a dough, which will be palatable when baked. When water is added to wheat flour, the proteins begin to absorb it. During mixing of the dough, individual protein chains join together, resulting in a strong, elastic network in the dough mass (gluten development); this is important as it traps the carbon dioxide gas generated by the fermenting yeast, which allows the dough to expand and eventually results in the soft texture of the final product. The water absorption capacity of the flour and the amount of protein in the flour (its 'strength') will significantly affect the quality of the final product, as will the amount of mixing. Other important factors are the degree of enzyme activity in the flour, time and temperature of mixing and proving (the time the final dough is left to stand before baking) and the level of salt added. Salt is a key ingredient in both the flavour of bread and its physical formation. Flour pretreatments and the incorporation of other ingredients such as ascorbic acid (vitamin C), fats and emulsifiers also contribute to the production of a wide variety of bread types. Thus, the production of bread involves the individual process steps of mixing, blending, fermenting and baking.

Cereals products - pastry

There are just three main ingredients in pastry - flour, fat and water - although many other minor ingredients can be added, such as sugar for sweet pastries. There is a huge variety of pastry products available, ranging from very soft sweet 'Danish' types to the very hard crusty pastry used to make pork pies. The variation is due to the ingredients used and the way in which they are put together. Low-protein flours are ideal for pastries for jam tarts and similar products, whereas savoury pies require flour with much higher levels of protein; in pork pies, the increased amount of flour, and the fact that the paste is boiled, are the most important features. A variety of fats can be used in the creation of pastry doughs, from animal lard to vegetable margarines. However, in the creation of puff pastry, fat is introduced after the formation of the dough and acts to keep the dough layers apart. It prevents the dough from agglomerating, is impervious to water and so creates lift and the characteristic flakiness of the product.

Cereals products - cakes and sponges

Traditionally, cakes contain fat and are aerated chemically (e.g. by the addition of baking powder), whereas sponges have less fat and are aerated mechanically (e.g. by whisking). Both usually also contain egg (although vegan alternatives are available). A cake recipe is a balance between three groups of ingredients: flour and egg, which provide the starch and protein that give cakes their mechanical strength; sugar, fat and baking powder which open the texture and provide lightness; and liquid ingredients such as milk and water, which make the texture heavier and denser. Each of the ingredients has a variety of functional properties. Sugar and egg are worth mentioning specifically, as they exemplify the highly complex nature of ingredient interactions in these and other products.

The main sugar used in cake making is sucrose. As well as its sweetening properties, it acts as a bulking agent, stabilising and controlling the viscosity of the foam/batter system, and it raises the temperature of starch gelatinisation and egg protein coagulation, which has a major impact on the structure of the cake. It also acts as a crumb tenderiser and as a humectant, controlling the water activity of the product; thus, cakes appear and taste moist, but microbiological growth is inhibited because of the lack of available water. Alternative sweeteners will have a different combination of properties and will result in cakes with different characteristics.

Eggs also have a variety of functions in cakes, being involved in emulsification, aeration, structure formation and moistening effects. Egg white contains about 13 different proteins, which themselves have a wide range of properties, including binding to starch granules; this gives the cake crumb its viscoelastic characteristics. As with sucrose, finding a direct replacement for egg is not easy, and requires significant overall reformulation to produce a comparable product.

Could these established foods pass the scrutiny facing today's novel foods and ingredients?

Some food crops are derived from wild ancestors which naturally produce toxicants, and the cultivated forms still produce low levels of the toxicant. However, because they have a safe history of use, and because the potential for toxicant production is taken into account during processing, we accept the foods as a safe and normal part of our diet. Despite their history of safe use, the question remains as to whether they would be accepted if introduced as novel foods today. Other products are consumed primarily because they have specific physiological effects - alcoholic beverages, coffee and tea, for example. For these products, the same question applies.

Food/beverage	Issue
Potato	Glycoalkaloids (toxin)
Peanut	Potentially fatal allergy in some individuals
Kidney bean	Lectins (toxin)
Rhubarb	Oxalic acid (toxin)
Alcohol	Chronic liver disease/addiction/drunkenness
Tea/coffee	Stimulant/sleep disturbance/addiction
Puffer fish*	highly deadly toxin
Onions	Severe lacrimatory effects
Curries/chilies	Sweating/mouth burning sensation

*(*not generally consumed in the UK)*

References:

Emsley. J. (1994) The consumer's good chemical guide. W.H. Freeman.

Emsley. J. and Fell, P. (1999) Was it something you ate? Food Intolerance. Oxford University Press. ISBN: 0 19 850443 8.

Bedford, L.V. (1986) Leguminous crops and their pulse products: a guide for the processing industry. Technical Bulletin No. 59. CCFRA.

Noah, N.D., *et al*. (1980) Food poisoning from raw red kidney beans. British Medical Journal, 19 July. 236-237.

Smith, D.B. *et al*. (1996) Potato glycoalkaloids: some unanswered questions. Trends in Food Science and Technology, **7**(4) 126-131.

2.3 Conclusions

New ways of processing foods have been developed since the beginning of mankind, but the last hundred years has probably seen more developments than the previous ten thousand. The perceptions of hazards and risks by today's consumer, and associated media attention mean that new products and processes are far more closely scrutinised than they ever used to be. There is no reason to suppose that the new processes are inherently more hazardous than more traditional ones or that the public are being used in a mass trial, as is sometimes portrayed. There is no doubt that if canned foods were being developed in today's climate, or if the industry was seeking to develop an edible vegetable from a poisonous plant (as has been achieved with the potato), they would have a hard time gaining acceptance.

3. FOOD PACKAGING ISSUES

Food packaging has several roles to play in the delivery of food to the consumer. It must protect the food from microbial, chemical and physical contamination, as well as keeping it in its desired physical state. It must inform the purchaser of what is in the food, and there are legal obligations to do this in most cases. It can also act as a medium for advertising the product and persuading the consumer to chose it rather than competitor products.

However, the packaging must not adversely affect the safety or quality of the product, and increasing environmental awareness and constraints on the disposal of the packaging mean that excess, unnecessary packaging is avoided. It must also be easy for the customer to use. Packaging can be expensive, adding a significant amount to the cost of the final product to the consumer - another reason for avoiding unnecessary usage.

3.1 Types of packaging

The type of packaging used will depend on the nature of the food and the processing and storage conditions to which it will be subjected. Also of importance are the types of functions that the packaging will have to fulfil (for example, does it need to be permeable to allow gas exchange, as in modified atmosphere packaging of fresh fruit or vegetables, or does it have a role to play in the cooking of the product, as in a microwaveable pack). Often the shape of the packaging (and possibly therefore the type used) will be constrained by the food in question though in some cases the reverse is true. A sandwich manufacturer, for example, will bulk-buy right-angled triangular sandwich cases and will require bread of closely specified shape (e.g. rectangular not domed) and size.

Table 9 - Types of packaging used in chilled foods

Aluminium foil	Plastics, e.g.:
Cardboard	polypropylene
Cellulose	polystyrene
Cellulose fibre	polyvinyl chloride
Glass	polyethylene - high or low density
Natural casings	polyethylene terephthalate
Paper	cellulose acetate
Metallised board	ethylene-vinyl acetate
Metallised film	
Steel	

Reference:

Day, B.P.F. (2000) Chilled food packaging. In: Chilled Foods: A Comprehensive Guide (Eds M. Stringer and C. Dennis). Woodhead Publishing.

Glass

Glass is one of the oldest packaging materials used for food, and is suitable for a range of products. It can be hot-filled (e.g. with jam or sauces), thermally processed, and stored chilled (e.g. milk). Its chemical inertness means that any type of product can be packed in glass, and its clarity means that it may present the food to its best advantage. Problems with light-mediated deterioration of some food types can be reduced by the use of coloured glass (e.g. wine in green or brown bottles).

Glass is not suitable for storing products in the frozen state, as it is likely to crack on freezing (as water, the main component of most food, expands as it freezes). Indeed, its susceptibility to breakage is the main drawback in its use.

Safe design of glass containers

Glass containers can be designed in a wide variety of shapes and sizes corresponding to both functional and marketing requirements. Despite the range of shape possibilities, the design must work within constraints for strength and safety. Carbonated beverages, for example, will always have circular or nearly circular cross sections, to maximise resistance to internal pressure. There will also be due notice taken of the resulting centre of gravity and size of the base, to ensure good stability: for this reason, the base will always be slightly concave. The type of glass used not only has to be compatible with the food it will contain and the process to which it will be subjected, but also with the closure to be used; much care is taken to ensure that the combination of these four elements does not result in an unsafe end-product.

Strength can be enhanced by avoiding high shoulders, square body sections and single container-to-container contact points. Special attention is also given to the sidewall contact areas. Generous contact areas reduce impact damage and label protection panels remove the risk of label scuffing.

Modern three-dimensional computer-aided design systems are now commonly used in glassworks design centres. These are particularly valuable in ensuring coherent and reproducible geometry where complex and non-rounded shapes are involved. When linked to computer-aided mould manufacture, the same 3D geometry is used to construct the mould cavities and the need for patterns, templates and models is completely eliminated

Reference:

Rose, D. and Gaze, R.R. (1998) Safety packing of food and drink in glass containers: guidelines for good manufacturing practice. CCFRA Guideline No. 18.

Metal

The two metals commonly used as the major component of food packaging are steel and aluminium. Relatively thick steel-based materials are used in canned foods, while thinner aluminium cans are used for soft drinks, and very thin films are used as foil laminates (e.g. as peel-off lids). Metals are very good light and oxygen barriers (although the barrier to oxygen will rely on a hermetic seal being formed in the package), and they can stand extremes of temperature, from sterilisation temperatures (typically 121°C) to below 0°C.

Metals can react with food, and lacquers and laminates are often used as can linings to prevent deterioration of the product. These linings are frequently used for heat-processed canned foods, which usually have a shelf-life measured in years, and which would otherwise be susceptible to internal corrosion. In some cases, the cans are designed to allow tin to migrate into the food, to maintain its colour and prevent other unwanted changes.

Preventing can defects

As cans are usually designed for prolonged storage of foods, it is important to prevent can damage, which would compromise the safety of the food. There are many stages during the lifetime of the can that could potentially give rise to a problem. These include:

- manufacturing operations: faults to either the can or the lid;
- receipt, storage and conveying of the empty can to the filling area: corrosion from rain, and physical damage because of poor loading, palleting or driving of fork-lift truck
- can filling operations: physical damage to the can or incorrect alignment or sealing of the lid
- can transport prior to retorting (sterilisation): physical damage through poor handling during transport or loading into the retort
- processing operations: corrosion because of poor quality sterilising water or excess oxygen in the retort, or too much chlorine in the cooling water
- post-processing operations: corrosion and physical damage during cooling and transport of the final cooled product respectively

Reference:

Thorpe, R.H. (1994) Guidelines on the prevention of visible can defects. Technical Manual No. 37. CCFRA.

Plastics

The use of plastic packaging is now widespread in the food industry. Different plastics can be formulated to have an array of different physical properties. As a result, different plastics can be hot-filled, sterilised at high temperature, or kept under chilled or frozen conditions. Plastic is also very amenable to being formed

The emergence of semi-rigid plastic packaging of food

The production of food in semi-rigid containers, that require a heat treatment to achieve long-term stability of the food under ambient storage conditions, has emerged over the past 15 years, as an addition to the traditional can, offering extra benefits of presentation and convenience for premium quality food products. Typical containers are a rigid pot or tray with a flexible metal or plastic foil lid. Modern automatic processing systems have greatly enhanced the uses to which manufacturers can put this type of package. The containers are manufactured from combinations of thermoplastics, sometimes with other materials added to give additional properties. A detailed technical performance specification will be required to ensure that the package is suitable for the food it contains and the process to which it will be subjected.

As with canning systems, the packaging has to be protected from damage which could give rise to loss of container integrity and subsequent microbial contamination; for the same reason, the filling and sealing operations have to be carefully controlled, and the fill weight is particularly important. After heat processing, the containers have to be rapidly cooled to about 45°C, and the pressure within them has to be controlled to prevent undue stress on the lid seams.

All-plastic containers have found a major niche for products requiring microwave reheating, as metal containers are not suitable for use in microwave ovens. This has extended the role of this type of packaging to chilled and frozen ready meals, requiring only a pasteurisation or cooking step, rather than a full sterilisation process.

Reference:

Campbell, A.J. (1991) The shelf-stable packaging of thermally processed foods in semi-rigid plastic barrier containers. Technical Manual No. 31. CCFRA.

into different shapes and thicknesses, such as bottle and jars, pouches, trays, closures and laminate wrapping material, adding to its wide range of functionality.

Constituents of plastics can be transferred into food (called migration), and the plastics have to be formulated to minimise this. There are specific migration limits set in EU and UK legislation (See The Plastic Materials and Articles in Contact with Food Regulations 1998, SI 1376 as amended).

Paper and board

These are probably the oldest forms of food packaging still in regular use in the industry. As well as being used as structural components of a wide array of packaging materials, especially for dry and ambient-stable foods, paper is used as an element in most other types of packaging, principally for labelling purposes.

With suitable coating, paperboards can withstand a range of temperatures and their stiffness provides protection to delicate contents. Paperboard is also used in combination with plastic laminates; it is an excellent light barrier, but its porosity means that gases can pass freely through it.

3.2 Packaging functions

Barrier properties

Packaging helps prevent physical damage to food. It also functions as a barrier to micro-organisms, extraneous matter, oxygen and other gases, moisture, and light, and can also interact with flavour components of the food.

In many long-life foods, the virtual exclusion of oxygen is desirable, to prevent the occurrence of rancidity and other undesirable chemical changes. In other products, such as packaged fresh meat, a high level of oxygen may be required inside the packaging to maintain the desired colour.

Controlling moisture levels in foods is very important, generally from a quality perspective. Dry products such as crisps and biscuits can gain moisture and become texturally unacceptable, while bakery goods may become hard. Other products will dry out and also become unacceptable, if not correctly packaged.

Many foods will deteriorate if left in the light, especially in the presence of oxygen (e.g. wine and butter). Opaque or semi-opaque packaging will help to prevent this.

Interactive packaging

Packaging can be used to maintain or improve the quality of food in an active way. Canned acidic products such as tomatoes can be packed in unlacquered cans, allowing metal ions to react with the product to create the characteristic flavour. Tin coated onto the inside of the can may be desirable as it rapidly eliminates oxygen, which would otherwise cause deterioration of the product. The appearance of clear fruit juices is improved or maintained by the presence of tin. However, tin levels in the product are closely monitored and are tightly controlled by legislation.

In modified atmosphere packaging (MAP), the package is flushed with an artificial gas composition and sealed. The gas composition is chosen to optimise the keeping quality of the food. In many cases, the packaging will be impermeable, to maintain the artificial gas mixture for as long as possible. However, for fruit and vegetable products, which are still respiring, a material with a controlled rate of permeation is usually chosen.

Oxygen scavengers are sometimes incorporated into products; these usually take the form of sachets of oxygen-absorbing chemicals, but there are now several developments where the scavenger is being incorporated into the packaging itself.

Microwave packaging is now designed so that more even heating of the food can be achieved. Microwave receptor boards are metallised films attached to a carrier film or board; these absorb microwave energy to generate heat.

Tamper evidence

Tamper evident packaging is now commonplace, following a run of tampering incidents in the late 1980s. Initial effort also focussed on tamper-resistant packaging, but this is difficult to achieve, and tamper evidence has apparently been quite successful in limiting the amount of malicious contamination of food after packaging. Much packaging is inherently tamper-evident, but glass jars, bottles, and other containers may be opened and reclosed without any obvious signs being visible. In these cases, plastic overwrapping or caps and lids which pop up or break a vacuum can be used to indicate if the package has been opened.

3.3 Environmental aspects

The growing awareness of environmental issues and the need to control the growing volume of packaging has resulted in both legislative and marketing/economic initiatives to encourage a reduction in the amount of packaging being used and the reuse or recycling of that which is used.

Table 10 - Waste management hierarchy

Ranking	Treatment/Disposal	Definition
1	Waste minimisation	Limiting the amount of waste produced at source
2	Reuse	Reusing the material for its original purpose
3	Recycling	Involves some reprocessing of the material before it can be used again. Also includes: ◆ Composting ◆ Landspreading ◆ Energy from waste
4	Disposal	Actual removal of the material from the waste stream. Includes: ◆ Incineration without energy recovery ◆ Landfill

The waste hierarchy (Table 10), as encapsulated by UK government white papers (e.g. DETR, 2000), has evolved to act as a guide to the possible disposal options for a particular waste and emphasise their relative priority (ranking) with regards to environmental friendliness. While the higher level options are the preferred, some wastes still have to be disposed of by incineration or by depositing in landfills, due to their nature, e.g. the waste may be too hazardous to be recycled and must be incinerated. These topics are dealt with in detail in Cybulska (2000).

4. FOOD SAFETY HAZARDS

Food safety hazards come in three basic categories: microbial, physical (e.g. foreign bodies such as glass), and chemical. Both chemicals and micro-organisms of concern can occur naturally in the food raw material and all three types of hazard can potentially gain access to food during harvesting, processing and production. Ways of preventing or minimising these hazards are discussed in Chapter 6. This chapter looks at some of the most significant areas of concern.

4.1 Micro-organisms of concern

Pathogenic micro-organisms are the major food safety concern for the industry. As they are generally undetectable by the unaided human senses (i.e. they do not usually cause colour changes or produce off-flavours or taints in the food) and they are capable of rapid growth under favourable storage conditions, much time and effort is spent in preventing, controlling and/or eliminating them. The vast majority of outbreaks of food-related illness are due to pathogenic micro-organisms rather than chemical or physical contaminants. Brief details of some of the most important organisms are given below.

Escherichia coli

E. coli is a ubiquitous inhabitant of mammalian intestines including humans. There are hundreds of different strains and variants, the vast majority of which are not harmful. However, some are not benign, such as the verocytotoxin-producing group, of which *E. coli* O157 is a member. *E. coli* O157 is primarily present in cattle, which are unaffected by it, but it is particularly dangerous to people on several accounts. Doses of just a few cells (possibly as low as 10) can result in illness. The organism is resistant to acid (low pH), and so can tolerate conditions in the stomach, which may contribute to its low infective dose. The toxin it produces is highly potent - causing anything from mild diarrhoea to serious urinary and gastrointestinal complications, including internal bleeding. An outbreak in Japan in the mid-1990s affected about 10,000 people and killed 9 people, while the outbreak in Lanarkshire, Scotland in 1997 caused illness in around 500 people and resulted in 22 deaths. There have also been several serious outbreaks in the USA. In all cases, those surviving the infection remain at risk of long-term kidney malfunction. *E. coli* O157 is mainly associated with meat products such as minced meat and burgers, but apple products, such as unfermented cider (common in the USA) are also at risk, via contamination of the apples with manure. Because of its low infectivity threshold, there have been instances of other products causing illness as a result of cross-contamination.

The organism is effectively destroyed by heating at 70°C for 2 minutes. For more information on this organism see Bell and Kyriakides (1998a).

Clostridium botulinum

Although this organism is the most notorious in the food industry, producing a highly potent neurotoxin, it is to the immense credit of the industry that incidents of botulinum food poisoning are very rare events in the UK. *C. botulinum* is an obligate anaerobe - it requires an absence of oxygen to grow and produce toxin. The toxin attacks the central nervous system, causing paralysis of breathing amongst other effects. The mortality rate in infected individuals may be as high as 50%. Historically it has been mainly of concern to the canned food industry (where the food is present in an hermetically sealed container, most parts of which are anaerobic), although any food in which anaerobic areas exist could be at risk. Recently, vacuum-packed foods have been highlighted as being potentially at risk, and special precautions are required for these products (Betts, 1996). The products are perfectly safe if the correct procedures are implemented in their production.

The main characteristic of *C. botulinum* is that it produces heat-resistant spores under adverse conditions. When conditions are favourable, these spores can germinate, grow and produce the highly potent neurotoxin. Canned foods can provide that ideal environment, so the organism must be eradicated in the canning process. These processes are based on the mathematical principle of reducing the chances of one spore surviving in a can to 1 in 1 million million. The basic process for this is heating all parts of the can to 121°C for 3 minutes (a 'botulinum cook'), although this time/temperature combination may need to be significantly altered depending on the food type.

Growth of most strains of *C. botulinum* is inhibited at refrigeration temperatures; however, psychrotrophic (cold-growing) strains do exist. This means that vacuum packed (or 'sous vide') products, which may not have received a sufficiently high level of heat process, can only be safely stored under strictly controlled chilled conditions for a short period of time. For further information on this micro-organism see Bell and Kyriakides (2000).

Listeria monocytogenes

The most important characteristics of this pathogen are its widespread natural occurrence and its ability to grow at refrigeration temperatures. It is essential that it is not present in foods which are not going to be cooked prior to consumption and, in the past, it has been of particular concern in the production of soft (unpasteurised) cheeses and pâtés. Infection with this organism can be fatal, especially to the young, old or immunocompromised. It is also a serious problem for pregnant women, as it can cause spontaneous abortion. Overall mortality rate of susceptible individuals can be as high as 30%.

The organism is destroyed by processes equivalent to 70°C for 2 minutes. Its growth is inhibited by mildly acidic conditions (below pH 5), but it is fairly salt-tolerant (up to 10%), and is somewhat resistant to drying. For further information on this micro-organism see Bell and Kyriakides (1998b).

Salmonella species

There are many *Salmonella* strains known to cause food poisoning. They are particularly associated with poultry and egg products but outbreaks have been associated with a wide range of foods from corned beef to salad vegetables. Because they can be present in chicken carcasses, it is essential that whole birds are properly cooked through, as *Salmonella,* if present, is likely to occur in the centre of the carcass. A temperature of 70°C for 2 minutes is sufficient to destroy *Salmonella*.

The '*Salmonella* in eggs' scare is widely seen as the first in a series of major UK food scares emerging from 1988 onwards. It followed a claim by a Government Junior Minister that UK eggs and poultry were widely contaminated with *Salmonella*. In the ensuing debate, particular concern was expressed over *Salmonella* Enteritidis Phage Type 4 (PT4) as it can contaminate eggs if it infects chicken reproductive tissues. The number of food poisoning cases associated with isolation of *Salmonella* Enteritidis from humans rose sharply during the late 1980s and 1990s - from around 500 isolates in 1981 to over 23,000 in 1997. However, there has been a significant decline recently, to about 10,000 in 1999 (Anon, 2000).

A two-pronged approach was adopted to tackle the problem of *Salmonella* in eggs and poultry. Firstly, compulsory testing of laying flocks was followed-up with slaughter of all flocks found to be infected. This led to the compulsory slaughter of nearly 400 flocks, totalling around 2 million birds, between March 1989 and February 1993. The second approach was to introduce measures to reduce contamination of poultry feed via a series of codes of practice. Typical measures included vaccination of flocks, adopting, implementing and documenting systems to ensure that raw materials and final feeds were of satisfactory bacteriological quality, and supporting this with appropriate analytical monitoring, personnel training and hygiene, and hygiene of premises and transport systems (MAFF, 1989).

Campylobacter

Although not as well-known to the general public as other micro-organisms, *Campylobacter* is by far the commonest cause of gastrointestinal infection in England and Wales (Stuart *et al*., 1997), and worldwide (Phillips, 1995) although few cases are linked to outbreaks (Pebody *et al.*, 1997). It has a low infective dose, but does not generally grow in foods, and is associated with food as a result of cross-contamination. It was not until the 1970s that *Campylobacter jejuni* was first recognised as a cause of foodborne diarrhoea, and assessing the true incidence of *Campylobacter* infection is difficult because of under-reporting. This is due to the fact that the illness caused by infection is often relatively mild. In some cases, infection may go almost unnoticed, but at the other end of the spectrum, abdominal pain caused by infection can be mistaken for acute appendicitis, and there may be a link with certain cases of Gullain-Barre syndrome (Phillips, 1995). This is a gradual degeneration of nervous function, which results in loss of mobility and sometimes loss of control of vital functions such as breathing.

Many of the cases are believed to originate from food, with strong circumstantial evidence linking illness caused by the organism with consumption of a particular foodstuff. For example, one particular outbreak of *Campylobacter jejuni* in a residential school was associated with milk bottle tops being pecked by birds (Stuart *et al.*, 1997).

Poultry is one of the most common foods associated with infection. *Campylobacter* is a microaerophilic thermophile - it likes warm conditions with small amounts of oxygen. As such, it will not generally multiply during normal storage conditions, especially refrigerated conditions. Like the other major vegetative pathogens, a process equivalent to 70°C for 2 minutes is sufficient to destroy the organism. However, undercooking of poultry carcasses might provide perfect conditions for its growth (Phillips, 1995).

Staphylococcus aureus

Poisoning from this organism results from the ingestion of a heat-resistant pre-formed toxin produced in contaminated food. *S. aureus* is notable for its ability to grow under fairly dry conditions (down to a water activity of 0.86) and for its tolerance to salt (up to 15-20%). However, it does not compete well with other organisms, and is therefore most often found in foods where other organisms are unable to grow (e.g. in high-salt foods such as hams) or in foods where other organisms have been excluded (e.g. cooked meats) but which *S. aureus* has managed to contaminate after processing. The most common source of such contamination is food handlers, as the organism may be found on the skin, particularly in the nose.

Bacillus cereus

This is another spore-forming organism, which exerts its food poisoning effect via the production of a toxin. While the vegetative cells are sensitive to heat, the spores can resist mild cooking regimes. The organism is quite often associated with starchy foods, especially cooked food held at warm temperatures, when the spores can germinate and the resulting cells multiply extremely rapidly. Eating food contaminated with the organism can result in a relatively mild diarrhoeal food poisoning, but if the cells have had the time to produce significant amounts of toxin, a more serious emetic food poisoning can occur. The latter has been associated with fried or boiled rice and has been termed 'Chinese takeaway' food poisoning. Most strains do not grow well below 10°C, so there is rarely a problem with adequately cooked, chilled products.

Other bacteria

Other organisms of significance as causes of food poisoning include: *Clostridium perfringens*, a spore-forming organism often associated with meat, soups and gravies; *Yersinia enterocolitica*, like *Listeria*, capable of growth at refrigeration temperatures; and *Vibrio* species, which are salt-loving, requiring 2-4% salt for optimal growth.

Protozoa

Cryptosporidium is a water-borne protozoan parasite (i.e. it is not a bacterium) that is of particular concern to the water industry. It is more resistant to chlorine disinfection regimes than the bacterial pathogens historically associated with drinking water. It forms resting, environmentally resistant stages called oocysts, and these can outgrow in the small intestine after drinking contaminated water. Children up to the age of 5 are particularly susceptible to the organism, and an ID50 (dose which causes illness in 50% of those exposed to it) of 100 oocysts has been suggested.

The organism is susceptible to heat, being destroyed after 5 minutes at 60°C, so water that is suspected of being contaminated can be rendered safe by boiling. It is also susceptible to desiccation, but is at least partially resistant to freezing down to -22°C.

Monitoring for this organism on a regular basis is not a straightforward matter, and monitoring for oocysts in treated water is currently only recommended under abnormal circumstances. Thus, the first indication of an outbreak of this organism in domestic water supply may come from a Health Authority that notices a cluster of incidents of illness. More details on the characteristics of *Cryptosporidium* are given by Dawson (1998, 2000).

Other pathogenic protoza include the waterborne *Giardia* and *Cyclospora* (which has been associated with soft fruit)

Viruses

Another area of growing concern is foodborne illness caused by viruses. These are extremely simple forms of life, consisting of a protein coat around a nucleic acid (RNA or DNA) core. They replicate by incorporating their RNA or DNA into that of their hosts. There are a wide variety of viruses, but only a few are of food safety concern. Most cannot survive outside of their host (e.g. on food), and none can multiply in food.

The incidence of food poisoning from viruses is fairly small, although symptoms can be quite severe. Outbreaks caused by viruses are characterised by incubation periods of 24-48 hours, a high attack rate and typical gastroenteritis symptoms of vomiting, diarrhoea and abdominal pain. In England and Wales, several outbreaks have been attributed to parvoviruses and the foods involved are usually seafoods, particularly cockles, but also mussels and oysters. Filter-feeding molluscs probably accumulate the viruses from organic matter in polluted waters, and the viruses survive subsequent cleaning and heat treatment. Norwalk viruses have also been implicated in outbreaks involving cooked meats, the most likely source of contamination being food handlers.

Bovine Spongiform Encephalopathy (BSE)

It is worth considering the BSE problem as an issue separate from standard microbiological problems, as it appears to be mediated through an infectious protein - a prion. It is almost certainly the food scare which has had the single biggest impact on public confidence in food safety in the UK.

BSE in cattle, along with scrapie in sheep and CJD in humans, is an example of a transmissible spongiform encephalopathy (TSE). Although scrapie is known to have existed in the UK and other countries for over 200 years, the first case of BSE is believed to have been observed on a Sussex farm in December 1984. In 1992, when the disease reached its peak, there were 36,682 confirmed cases. The origin and transmission of BSE is still the subject of debate. One theory is that BSE has its origins in scrapie, and spread via scrapie and/or BSE-infected material present in cattle feed - possibly exacerbated by changes to the rendering process that eliminated some heat processing and solvent treatments that might otherwise have denatured the infective agent (prion). More recently it has been suggested that the disease arose by chance - as a spontaneous mutation in the prion gene of sheep or cattle in the 1970s - which then spread via contaminated animal feed.

The scale of the impact of BSE on consumer confidence is due to the suggested possible link with new variant Creutzfeld-Jakob disease (vCJD) - the human condition suggested to arise from consumption of BSE-infected beef. vCJD emerges through appalling and protracted symptoms, robbing victims of their independence, their dignity and ultimately their life. The lag between infection and emergence of symptoms is believed to be around 12-15 years (some suggest it might be as long as 40 years). The biology of the disease is poorly understood, there is no cure and relatively little can be done to alleviate the symptoms. Belief in the mechanism by which the disease is transmitted rests on circumstantial evidence, and even the best informed experts are still unable to predict the eventual scale of the problem.

The fact that CJD (including vCJD) currently remains an extremely rare condition - the number of cases in the UK has fluctuated around 80 per annum since 1997, which equates to just over one per million of the population - is, for many, far outweighed by the nature of the disease and the possibility that it might lie latent in an already infected population.

Early measures in the UK to control the BSE epidemic involved widespread culling of infected herds, restrictions on the age of cattle sent for slaughter, changes in abbatoir practic and the prohibition of feedstuffs containing ruminant material - the former was also aimed at directly reducing the risk to human health. The measures, albeit at enormous cost, have certainly had the desired effect on incidence of BSE in cattle in the UK; by 1998 the number of confirmed cases had fallen to 3,064 and the downward trend continued in 1999/2000. However, earlier official assurances as to the safety of beef and beef products in the wake of these measures were regarded with significant scepticism by many. Confidence was further undermined by the announcement on 20th March 1996, by the UK Health Minister, that the most likely explanation of the cause of 10 cases of the new form of CJD was exposure to BSE-infected material. Later that week, the Chairman of the Spongiform Encephalopathy Advisory Committee (SEAC) [to the UK government] was quoted as saying that perhaps half a million people in the UK were infected with CJD. On completion of the BSE enquiry in October 2000, projections of numbers of possible vCJD cases still ranged from tens to tens of thousands.

For the meat industry, the consequence of the 1996 announcements were profound: many stores and catering outlets stopped selling British beef and the EU's standing veterinary committee voted to impose an indefinite world-wide export ban on British beef and beef products. Although this ban has now been lifted, it is possible that the sale of British beef products will be affected for decades in certain sections of the world community. Meanwhile, cases of BSE are now emerging on mainland Europe.

References:

Adams, J. (1999) Cars, cholera, cows and contaminated land - virtual risk and the management of uncertainty. In: What Risk? Science, Politics ad Public Health. Ed: Bate, R. Butterworth Heinemann pp 295-304.

BSE Enquiry (2000) BSE Inquiry: the report. www.bse.org.uk

Craven, B.M. and Stewart, G.T. (1999) Public policy and public health: coping with potential disaster. In: What Risk? Science, Politics ad Public Health. Ed: Bate, R. Butterworth Heinemann pp 222-241.

Department of Health (2000) Monthly creutzfeld-Jakob disease statistics (October 2000). www.doh.gov.uk/cjd/stats/Oct00.htm

Ironside, J.W. (1999) nvCJD: exploring the limits of our understanding. Biologist **46** (4): 172-176.

IFST (1999) Position paper on bovine spongiform encephalopathy (BSE). June 1999 edition.

4.2 Chemical hazards

Food itself is made up entirely of chemicals. Other chemicals may be incorporated into food intentionally (as part of the process) or occasionally accidentally and may pose no risk at all. Others which can pose a hazard at high enough levels may unavoidably occur at very low levels. We are naturally exposed in all walks of life to low levels of chemicals that are toxic at higher concentrations. Providing that specific hazards are identified, understood and controlled - so that exposure to such chemicals remains low - there is negligible risk to health. [See Elmsley (1994) for more on the 'threshold' concept].

Many chemicals in food that do pose a food safety issue are not actually contaminants at all, but natural components of the food (Emsley and Fell, 1999). Some are of microbiological origin, and while being highly undesirable, are 'naturally' associated with the food. Ironically, modern analytical techniques, which mean that we can now detect the presence of extremely low levels of contaminants and take preventative or remedial action, have focussed attention on the whole subject of potentially harmful chemicals in food. The industry takes great care to minimise the possibility of any potentially harmful chemical getting into food. In addition, the government carries out regular surveys where there is known to be a potential risk to ensure that any exposure is below defined safety levels. One example is the monitoring of milk for dioxins, which are widespread in the environment, and can potentially accumulate in milk. The surveys have shown that, generally, dioxin levels are well within safety limits: they have also allowed rapid action to be taken on farms where higher levels have been detected, before they become a problem.

It is the safety of the final product which is of concern to the manufacturer, and so this section will include details of some of those chemicals of significance to the industry, both 'natural' and extraneous contaminants. It is not possible to include details of all naturally occurring toxicants: these are numerous, but the examples given provide an indication of the types of problems that the food processor has to consider.

Other chemicals that might have an impact on quality are covered elsewhere.

Mycotoxins

Mycotoxins are toxic compounds produced by fungi. There are many different mycotoxins, but only a small number of fungi have been shown to be a problem in food production and consumption. Fungal contamination and associated mycotoxin production is generally a result of poor storage conditions. Historically, ergotamines from contamination of wheat and rye with *Claviceps purpurea* were shown to have a causal relationship with human illness (in the Middle Ages, the resulting combination of convulsions and hallucinations were called St. Anthony's Fire), but there have been no known cases in the UK for over 100 years (MAFF, 1987).

The symptoms of mycotoxin poisoning vary from one toxin to another, but they are generally long-term. Some mycotoxins have been linked with liver and kidney cancer.

The mycotoxins of most current concern to the food industry include the aflatoxins, ochratoxin A and patulin, although there have been no proven cases of food-related illness associated with any of them (MAFF, 1993). As their effects are both cumulative and chronic, such a link would be difficult to make, but the industry takes great care to minimise their occurrence in food.

The aflatoxins are produced predominantly by two fungi, *Aspergillus flavus* and *Aspergillus parasiticus*. These fungi are a particular problem during the storage of nut and fig products, which may become infested if appropriate (cool and dry) storage conditions are not maintained. Maximum limits have been set in legislation for aflatoxin levels in these products. Aflatoxins are also known to be a problem in animal feed; as these may arise in milk from cattle fed contaminated feed, dairy products are routinely monitored to ensure that levels are acceptably low (MAFF, 1993).

Ochratoxin A was originally isolated from the fungus *Aspergillus ochraceus*, but has subsequently been found to be produced by a range of *Aspergillus* and *Penicillium* species. These species are primarily contaminants of coffee, cereals and cereal products, including animal feeds, and monitoring of relevant foods, such as wheat, maize, oats, and barley, is routinely carried out (MAFF, 1993).

Patulin is produced by a number of species of *Aspergillus* and *Penicillium* including *Penicillium expansum*. Although the latter is a common storage rot of a number of soft fruits, patulin has only been noted to any significant extent in apple juice and similar products. Adherence to good manufacturing practice regimes, specifically control by selection and storage of raw materials to prevent growth, is an effective way of reducing this mycotoxin to acceptable levels.

Glycoalkaloids in potatoes

Glycoalkaloids are naturally present in potatoes and can result in gastrointestinal disorders if ingested in sufficient quantities. When potatoes are exposed to the light, they are prone to the formation of increased levels of glycoalkaloids near the surface. At the same time, they also produce chlorophyll, which causes the greening effect. The green pigment itself is not toxic, but it does indicate that toxic glycoalkaloids might also have been produced. Other stress factors that can induce glycoalkaloid formation include physical damage, pre-harvest stress such as waterlogging or drought, and sprouting (Smith *et al*., 1996). Some potato varieties contain higher glycoalkaloid levels than others. As the toxins are concentrated near the surface, larger potatoes will have relatively lower amounts than smaller potatoes, and peeling will significantly reduce levels. Application of HACCP (see Chapter 6.1) to growth, harvesting, storage and processing of potatoes will also minimise the amount of glycoalkaloid produced (Smith *et al*., 1996).

Kidney bean lectins

Lectins are natural components of raw kidney beans and are toxic. They are proteins and probably act via their ability to bind to sugar molecules on the cell surface of the intestinal wall and block absorption. The net effect can be nausea, vomiting and diarrhoea. In some cases, hospitalisation and intravenous infusion may be required (Noah *et al*., 1980). The lectins therefore need to be destroyed before the beans are eaten. This can be achieved by soaking overnight, discarding the water and then vigorously boiling the beans for about 10 minutes. It is important to boil vigorously, as gentle warming to sub-boiling temperatures has been shown to increase activity of the bean lectin. The type of process achieved in the canning of kidney beans is more than sufficient to destroy lectin activity.

Oxalic acid in rhubarb

It is widely known that rhubarb leaves are toxic. However, it is less widely known that the toxic component, oxalic acid, is also present at relatively high levels in the edible stalk. Whereas eating rhubarb leaves is not recommended (and there are historical reports of people dying after doing this), consuming the stalks is not generally considered to pose a problem. The occurrence of a small piece of leaf in a can of rhubarb, whilst undesirable, is also not a health risk.

Cyanogenic glycosides in fruit

Many plant materials naturally contain cyanogenic glycosides. Basically, these are sugar complexes that have the potential to release cyanide under certain circumstances. In many stone fruits, for example, the stone itself contains cyanogenic glycosides. In the raw fruit, they are not a problem as the substrate is physically separated from the enzymes that can release the cyanide. In the canning of stone fruit, the heat process will destroy this physical separation. The canner has to ensure that the heat process applied is sufficient to inactivate the enzyme and so prevent the possibility of cyanide being formed, but not too extreme that the product goes mushy (which soft fruits are prone to do).

Pesticides

The use of pesticides on fruit and vegetable crops is very closely controlled by law. Only approved pesticides may be used during the growing and storage of crops, and each approval, which is made on an individual formulation basis, often relates to a specific application rate on specific crops, and will specify a minimum time interval between application and harvest. Thus, in any given situation there are a relatively small number of pesticides that might be present. In addition to the control at the application stage, maximum residue levels (MRLs) are often set which reflect those levels that might be present following good practice in the application of the pesticide. These MRLs are set well below any threshold safety level, and so in the overwhelming number of situations, any remaining pesticides will not pose a health risk. However, the food processor will have to monitor levels to ensure that levels remain well within set limits.

Antibiotics

Meat processors need to be vigilant over the use of antibiotics in the production of their products. Appropriate use of certain antibiotics to counteract disease is allowed, but some formulations have growth-promoting and other effects and their inappropriate use could contribute to the development of resistant microbial populations. In a recent review (MAFF, 1998), the extent of antimicrobial resistance in the food chain has been summarised. Antimicrobial resistance in both the normal human microflora and in human pathogens has been well documented, as has such resistance in animals treated with the antimicrobials. The transfer of resistant bacteria from animals to man via food has also been documented, although the length of time these populations persist in an individual is not clear.

Heavy metals

Metals such as lead, arsenic and cadmium can occur in a variety of foodstuffs. They arise indirectly in foodstuffs from the environment - e.g. they are in the soil that the crop is growing in, or in the grass that the cow is eating or in the water in which the fish is living. As such, they cannot be 'removed' during processing and so control of raw material quality is the only mechanism for ensuring that levels do not become unsafe. Maximum levels of lead and arsenic, for example, are set in legislation, and conformance with these regulations will ensure that there are no safety issues. For other metals, there have been recommendations published by various government bodies as to safe maximum daily intakes. Again, adherence to these recommendations, by monitoring levels when a potential problem has been identified, should negate any possible health risk.

Packaging migrants

There has been much debate recently over components used in the production of plastic, which are widely used in packaging and handling food throughout the industry. In the UK, the 'Plastic Materials and Articles in Contact with Food Regulations' define what monomers, plasticisers and other components can be used in the manufacture of plastic food-contact materials. In addition, the general 'Materials and Articles in Contact with Food Regulations' state that components of the material shall not migrate into the food to an extent that would pose a health risk or deleteriously affect the quality of the food.

Dioxins and polychlorinated biphenyls (PCBs)

These chlorine-containing compounds, which are structurally quite closely related, are formed in variable amounts in many industrial chemical processes and combustion reactions. Of the 200 or so dioxins known, about 19 are believed to be toxic. One of them has been quoted as being the most toxic chemical known (Emsley, 1994).

If present, dioxins and PCBs are most likely to be found in fats and oils, and products containing or derived from them. In Belgium in 1999, a consignment of oil destined for the manufacture of animal feeds was contaminated with one for industrial (non-food) purposes; the latter contained significant levels of dioxins. This set off a trail of events which led to the closure of abattoirs in and around Belgium and the removal from shops of a wide range of meat and dairy products of Belgian origin. The delay by the Belgium authorities in notifying the food industry meant that audit trails to trace the origin of particular food products were often inconclusive and, as a safety precaution, many unaffected products also had to be withdrawn.

Other types of chemical contaminants

There have been a number of other 'one-off' incidents of chemical contaminants in food. Some of these have been the result of genuine mistakes or accidents; others have resulted from malpractice.

The Spanish toxic oil syndrome incident illustrates how an issue which starts as a deliberate act of fraudulent adulteration can also compromise food safety and so undermine consumer confidence in both food authenticity and food safety. In this incident, a large volume of rapeseed oil had been denatured with aniline to downgrade it for industrial use. The oil was, however, refined, decolorized, deodorised and mixed with other oils before being packaged and labelled as olive oil, and illegally introduced on to the Spanish market (Arribas-Jimeno, 1982).

Over 20,000 people suffered health problems - many with symptoms as severe as respiratory failure and muscle wasting - and as many as 600 people are believed to have died as a result of consuming the oil. The evidence linking the outbreak to the oil was initially epidemiological, and it was only after several years and extensive scientific investigation, that the causative agent was reliably confirmed.

4.3 Foreign bodies

The variety of foreign body items that have been reportedly found in food over the years is considerable. Glass, stones, metal, and insects can occur, and items as bizarre as whole rodents have been claimed. The principles of HACCP (see Chapter 6.1) and allied monitoring techniques can substantially reduce the chances of foreign bodies being found in food. It is up to food producers to recognise what contaminants are a realistic threat to their product and take appropriate steps to prevent their occurrence. Apart from general hygiene considerations (e.g. preventing the entry of rodents or insects) there are specific detection mechanisms that can be incorporated into a process to identify and remove the types of objects that may be inherent in the raw materials being used. Considerable effort goes into preventing contamination of foods with foreign bodies. Campbell (1995) provides a major overview of foreign body types and detection methods. Brief details of those of most concern to the food industry are given below.

For any given product or process line, there is likely to be more than one realistic potential problem with foreign body contamination. The food processor, using a HACCP approach, will need to identify those that are a realistic threat and put in suitable preventative measures and allied detection procedures. This may involve consecutive use of more than one type of foreign body detection system.

Metals

Contamination with pieces of metal is highly undesirable. However, as metal machinery is used in the harvesting, conveying and processing of food, there is always the possibility of pieces (e.g. nails, screws, wire filaments, ballbearings) contaminating ingredients during processing. These need to be located and removed. Various metal detection systems can be employed, though no metal detection system will be able to detect all metal contaminants. In addition to absolute size limits for detection, the nature of the contaminant and its orientation are particularly important. Some metals are more easily detected than others: for example, mild steel, which has good magnetic and conductive properties, is much more easily detected than stainless steel, which is less conductive

Glass

This is probably the most contentious of foreign body contaminants. Slivers of glass in a food can be highly dangerous, as they could cause severe internal lacerations and bleeding. Very small fragments of glass are very difficult to detect, and so very strict processing procedures need to be in place to restrict the chances of glass getting into food. In simple terms, this means prohibiting glass in food processing areas wherever possible, covering up lighting (which has to use glass) with suitable transparent shields, and stopping production and quarantining product if a breakage does occur in the vicinity of the product. Rose and Gaze (1998) describe the many procedures adopted and the precautions that have to be taken for products packed in glass. These range from adoption of best practice in handling glass packaging (e.g. eliminating glass completely from the production area) to using x-ray techniques to detect contamination.

It is worth mentioning that many innocuous substances in food are often mistaken for glass. Struvite (crystals of magnesium ammonium phosphate) is one example, being periodically formed and reported in canned fish and shellfish.

Foreign body identification

A foreign body can be defined as any piece of undesirable solid material found in food, ranging from pieces of bone or fruit stalk to fragments of glass or metal. Clearly, foreign bodies are of potential concern to all food companies and can have serious consequences for a company. At the very least it might undermine consumer confidence and tarnish brand image, but in more serious cases it could destroy a business or have knock-on effects for the entire sector. In dealing with a foreign body incident, the first step is to get a rapid and reliable identification of the foreign body in question.

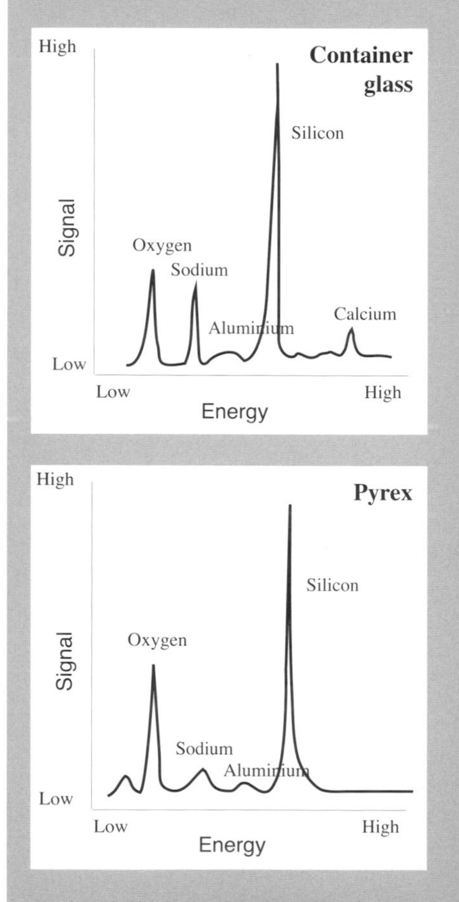

A good example is the identification of glass fragments. Glass samples can be 'typed' by x-ray microanalysis. This basically looks at the minerals present. Different types of glass contain different minerals and so display different profiles. Samples can be compared to a reference library to quickly identify the type of glass involved. This is vital in establishing the nature of glass complaints, many of which originate in the consumer's home from items like chipped Pyrex bowls. In these instances the company can be confident that the glass has not entered the food production chain, and can act accordingly. Where it has entered during food production, remedial action can be taken. The x-ray microanalyses depicted here demonstrate how different types of glass can be distinguished. The same technique can also be used to type pieces of stone, or to distinguish between different types of stainless steel.

To complement these analyses of inorganic materials, the technology of FT-IR (Fourier Transform Infrared) microscopy can provide rapid and precise analysis of organic materials such as plastics. Each type of plastic gives a particular fingerprint with FT-IR, and this can provide a clear idea of the nature of the plastic. Where laminates are concerned, separate fingerprints can often be obtained for each material present, improving test reliability.

The analysis can be backed up with a range of other tests too, For example, microscopy can be used to determine the origin of fibres, and taxonomic keys to identify intact pieces of insects or other animals. It is often also possible to determine whether the foreign body has been processed with the food - which gives further clues as to the point at which it entered the product. For example, certain materials might absorb colour from the product or become distorted by a heat treatment. In fact, a significant proportion of foreign bodies have been found to have been introduced in the home - usually accidentally but sometimes deliberately to extort compensation.

A structured preventative approach to food safety - based on the principles of hazard analysis critical control points and enshrined in a sound quality management system - can prove highly reliable in minimising occurrences of foreign body incidents. A responsible manufacturer will then also use information arising from foreign body incidents to re-examine preventative procedures and, when an unpredicted incident occurs, perhaps introduce additional measures.

References:

Campbell, A.J. (1995). Guidelines for the prevention and control of foreign bodies in food. Guideline No. 5. Campden & Chorleywood Food Research Association.

Edwards, M.C. and Redpath, S.A. (1995). Identification of foreign bodies from food. Guideline No. 4. Campden & Chorleywood Food Research Association.

Edwards, M.C. and Fincher, C.H. (1999). The effect of food processing on foreign bodies: a case study on baking. Review No. 13. Campden & Chorleywood Food Research Association.

Ponzi, J. and Edwards, M.C. (2000). The effects of processing on foreign bodies. A case study with in-container heat processing. Review No. 16. Campden & Chorleywood Food Research Association.

Stones

These are also likely to come in with raw material. Depending on what the raw material is, there are effective ways in which they can be removed. There are many air and liquid separation systems that can be used to separate stones from product, which rely on the different densities of the materials. In products such as peas, in which stones may appear, optical devices, such as colour sorting machines, can also be used. These simply spot objects which are not the required green colour, and so will also remove discoloured peas.

Toxic berries

These are a particular problem when processing small fruits. Optical sorting machines may well be able to spot different coloured or shaped toxic berries, but if they are similar to the product, it may require an expert human eye to survey the conveyor belt.

4.4 Allergens

Allergens are not necessarily contaminants, nor are they toxins. They are chemicals (usually proteins) that the vast majority of the population consume without any ill effects, but which in some people cause reactions ranging from skin redness, irritation and swelling to anaphylaxis (which is potentially fatal). An individual could react to one type of allergen and not to others, or might react to a range of allergens; these are often structurally related. Allergic reactions are caused by a response of the immune system which triggers histamine release by the body (in an analogous way to the reaction to a nettle or bee sting). In exceptional circumstances, the amount of histamine released is extreme, causing a precipitous drop in blood pressure, which may result in the victim collapsing with one of several anaphylactic conditions.

There are a limited number of food types which are known to produce such severe reactions in a significant number of people. The most notorious are peanuts and tree nuts, but shellfish, milk proteins, egg protein and some seeds can also pose problems. Clearly, the food industry has to make every effort to ensure that potential sufferers know which products to avoid. If the allergen is known to be an ingredient in the food, it is a relatively simple job to label it as such. However, extreme care also has to be taken to minimise the chances of contamination of other products which do not contain the allergen. The steps taken will include effective cleaning of the production line, personnel hygiene, correct factory layout, airflow control, and, increasingly, nut-free production areas.

4.5 Conclusions

In all discussions of contamination, the nature and origin of food has to be clearly born in mind. Even the most stringent approach cannot guarantee the elimination of all contaminants all of the time. However, the presence of contaminants in a food that make it dangerous, unfit to eat or of unacceptable quality is an offence under the Food Safety Act in the UK. The only defence that the food processor has is that of 'due diligence' - i.e. that all reasonable precautions and all due diligence were taken to prevent the contamination from occurring. Clearly most food companies invest considerable time, resource and money in developing products, establishing brands and cultivating brand loyalty. A single contamination issue, even if harmless to the consumer, can undo this in a matter of just days.

5. FOOD QUALITY ATTRIBUTES

As well as the many food safety hazards discussed in the previous chapters, there are specific quality considerations that have to be addressed in the manufacture of foods. These range from microbial spoilage and chemical or physical deterioration of the product over time to the actual nature of the food at the time of sale and whether its appearance, taste, texture and content are acceptable to the consumer.

5.1 Quality deterioration

Food can deteriorate in a number of ways: milk may start to curdle in days; fatty foods will go rancid after a few months; and sponges, cakes and bread will go stale, dry or soggy after a while. The preservation techniques described in Chapter 2 are equally applicable to some spoilage and quality issues as they are to safety issues. Some typical spoilage issues that individual sectors of the industry have to face are briefly described below.

Staling

Staling is a consumer perception associated with the drying and firming of many types of bakery products; it may take many forms, depending on the nature of the product (Guy *et al.*, 1993). Although there are similarities in the changes that cause a staling effect in bread, cakes, pastries and biscuits, the exact nature of the changes will be different; however, the consumer still perceives the end result as staling.

There has been much work put into investigating the causes of bread staling over the years. As long ago as 1859 it was found that bread firming was not merely caused by a loss of water (Hoseney and Miller, 1998). The staling of bread is associated with a general loss of flavour and the development of a dry taste, an increase in firmness, and several chemical and structural changes in the starch, amongst others. Although these changes occur over approximately the same time-scale, they do not run exactly in parallel, and the perception of staleness by the consumer is probably from the combination of a number of factors. It also seems likely that changes occurring in protein structure, as well as moisture distribution, are involved (Russell, 1982).

Work with Madeira cakes indicated that staling was related to the firming and drying out of the crumb (Guy *et al.*, 1993); however, staling still occurred inside moisture-proof bags and was found to be at least partially due to the migration of moisture to the crust. The firming was thought to be associated with starch crystallisation, similar to that occurring in bread.

With the many factors involved in the staling of bakery products, there has been much work put into modifying the formulation of products, including changing the type of fats and sugars used in products and the addition of other ingredients, and into processing and storage procedures. It was found in the work by Guy *et al.* (1993) that the staling of Madeira cake was slowed below 5°C; however, refrigerated storage of other types of bakery products can result in accelerated staling compared with storage at room temperature.

Rancidity

Rancidity is a potential problem for all foods containing a high percentage of fat or oil. It can be associated with the oxidation of the fat, the release of free fatty acids from triglycerides (i.e. forms where the fatty acids are combined with glycerol), or the release of volatile compounds which have a strong or unappetising odour. As with staling, rancidity is a consumer perception, and is very much dependent on the nature of the food being consumed. The chemical changes mentioned may be extensive in a particular product, but may not be perceivable by the consumer, perhaps due to the presence of other strong flavours (e.g. spices). As such the changes are not a problem, as they present no safety hazard. In other situations, the changes may produce very undesirable flavour changes which severely limit the shelf-life of the food, e.g. in oily fish. In formulated foods, certain antioxidants - whether additives in their own right or as a natural part of an ingredient - can sometimes be used to slow down the onset of rancidity.

The useful life of frying oils is limited by fat oxidation and free fatty acid release; this can be reduced by the removal of food debris, which promote oil degradation, and also by keeping the oil as dry as possible.

Enzymic reactions

A variety of enzymic reactions can pose a range of significant problems to food manufacturers. One of the most widespread is the browning of the surface of cut fresh fruit and vegetables, which is most obvious on white-fleshed products, such as apples, potatoes and aubergines. This is initiated by the enzyme polyphenol oxidase, which catalyses the conversion of naturally occurring monophenols to ortho-quinones, in the presence of oxygen. These then polymerise to form brown polyphenolic pigments. In the cells of the uncut fruit or vegetable, the enzyme and the monophenols are physically separated from each other, but the act of cutting breaks down the cell barriers, the products mix and browning quickly occurs. In various food processing operations, this browning is highly undesirable and antioxidants such as bisulphite and ascorbate (vitamin C) are used to prevent it. Research has also shown that inhibition of enzyme production (e.g. by genetic modification) is also possible; without the enzyme, the reaction chain is stopped at the first stage and browning does not occur.

The softening of fruit and vegetables is another enzyme-mediated spoilage problem. This occurs in a wide variety of products and is caused by internal enzymes (pectinases) breaking down the rigid cell wall structure, causing the fruit or vegetable to eventually lose its firmness and go mushy or pulpy. These and other enzymic reactions, being chemical in nature, can generally be slowed by storing the product at chill temperatures; however, this is not always the case - the blackening of banana skin is a polyphenol oxidase-catalysed event and this is accelerated at low temperatures because of the fragility of the cells of banana skins. These are broken down very quickly at chill temperatures, allowing enzyme and substrate to mix and rapidly cause browning.

Microbial spoilage

Microbial spoilage can take many forms, and is a major limiting factor in the storage of a wide range of food products. The spoilage can manifest itself in the physical appearance of microbial colonies on the food (e.g. in mouldy bread), or in a change in the taste or appearance of the product (e.g. sour or curdled milk).

Spoilage can be prevented or retarded by the use of one of the many preservation techniques described in Chapter 2, with freezing or chilling of unprocessed or minimally processed products being the major route. It is worth re-emphasising that spoiled food is not necessarily unsafe to eat.

There are many different organisms responsible for food spoilage, but the spoilage of individual foods is often associated with one or a small number of specific organisms. Bread is subject to spoilage by moulds, and also by *Bacillus subtilis*, a bacterium that gives rise to 'rope' (hence the term 'ropey' for anything that is sub-standard). Pickles and sauces are susceptible to spoilage by acid-tolerant *Lactobacillus* species, yeasts, and the mould *Moniliella acetobutans*, which can not only tolerate acetic acid, but can metabolize it as well.

There is a whole range of bacteria associated with milk and dairy products; lactic acid bacteria cause souring by fermenting lactose (milk sugar) to lactic acid; *Escherichia* and *Enterobacter* species give rise to off-odours and can cause 'blowing' in cheese-making - which results in a large number of gas holes in the cheese; butyric acid-producing *Clostridium* species can also have the same effect. Enzymes produced by bacteria can also cause problems in dairy products - *Bacillus* species produce protein and fat-metabolizing enzymes which cause sweet curdling and give rise to 'bitty cream' (where the cream separates into small semi-solid globules) (Ranken *et al.*, 1997).

Cuts of meat are also prone to surface spoilage by a mixture of micro-organisms, which results in a slimy or sticky feel to the meat. The surface of fish normally harbours high levels of bacteria, with *Pseudomonas* species predominating; it is these that cause the main odour and flavour changes associated with spoiled fish (Ranken *et al.*, 1997).

Moisture migration

The movement of water from one part of a product to another may cause the product to deteriorate. This may involve pastry going soggy as water leaches out of the contents of a pie or pasty, or a sponge absorbing moisture from a filling. Depending on the type of product, different approaches can be taken to prevent or delay these effects; having a water-repelling layer such as butter is one possibility.

5.2 Consumer-driven quality issues

The primary producers of food in the UK have faced a series of new consumer demands over many years. These probably reflect a logical sequence in the development of ever-greater demands from the affluent consumer and are mirrored across Europe. It should be noted, however, that as new demands are added, previously delivered requirements frequently remain so that the list continues to grow (Hall and Jones, 1999).

- The initial concern was with getting sufficient food to ward off hunger and starvation.
- Having the physical needs satisfied, the consumer became concerned about whether the food was as safe and nutritious as it could be.
- With a safe and nutritious food supply, the consumer's attention turned to choice and the desire for greater selection and year-round supplies of previously seasonal produce.
- Price then became the primary target and effort was focussed on ways of increasing yields and reducing production costs.
- With all of these demands met, at least in part, the consumer looked for consistently high quality products, which led to increased variety and choice. Much of the research in food production over the past ten to fifteen years has been on quality.
- The modern consumer has plenty of food (some would say too much), it is generally safe (contrary to what some believe), the cost is kept as low as possible by efficient production practices, the quality is consistently good and

most products are available throughout the year. Consumer attention is now turning to how the food is being produced and what impact the method has on social and environmental welfare. Perhaps the clearest examples of this are in the rise of the organic products market (see Chapter 1.4.1) and in fair trading initiatives.

Flavour and texture

The emergence of the chilled ready meal market was largely as a result of the consumer's desire for convenience with flavour and texture characteristics more akin to fresh products. Today's consumers have a very wide range of desires for different food characteristics, and the growth in the ethnic food market has been significant. In addition to the more traditional liking for 'Italian', 'Indian' and 'Chinese' food, there is now a desire for more specific and authentic dishes, such as Thai and various West Indian recipes, as shown in Figure 4.

Figure 4 - 1999 Ethnic styles within chilled readymeals, recipe dishes and added value products

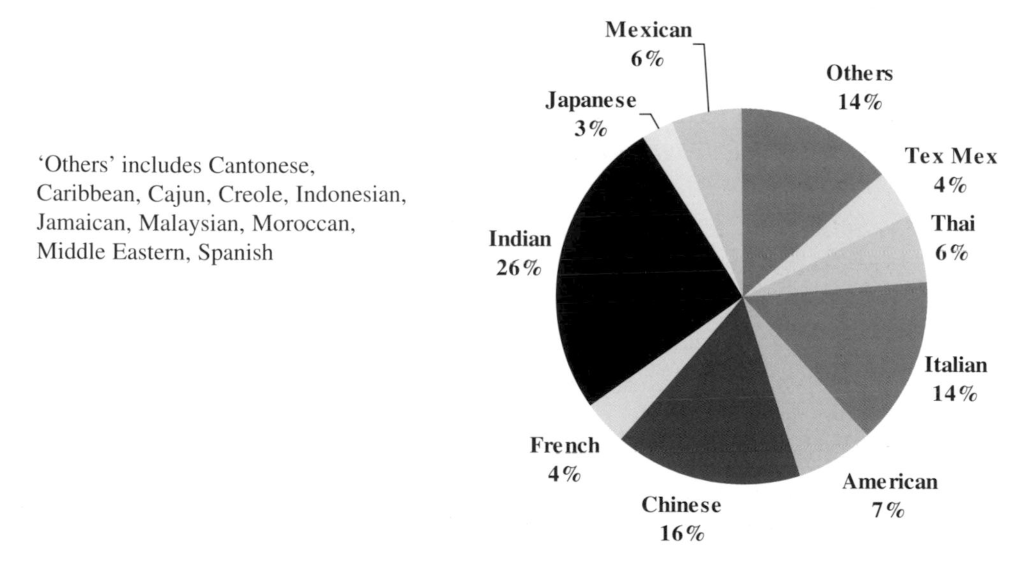

Nutrition issues

One of the major driving forces behind new product development is to provide the consumer with foods that they enjoy eating and which they feel is more nutritious in some way than a previous version. There has been a steady supply of low-calorie formulations since the 1960s, and these continue. However, the focus more recently has been for low- or reduced-fat products, to the extent that even some existing confectionery products have been advertised as '85% fat free' (i.e. the product still contains 15% fat). The UK government has recently issued guidance which suggests that 'x% fat-free' claims should not be made, as they can be confusing to the consumer. In addition to total fat, the current belief that monounsaturated fats are beneficial in comparison with polyunsaturated and saturated fats has had an effect on product formulation.

The level of added salt in food has also been a source of concern to some for many years, because of the risk of hypertension and associated cardiovascular illnesses from increased body sodium levels. Some reduced- and low-salt versions of foods are becoming available, and standard versions that are already low in salt are being highlighted.

Other compositional issues which are currently of concern to the consumer are sugar levels, with low- and reduced-sugar products emerging (especially for soft drinks), and dietary fibre levels, with many high-fibre products now being developed or marketed as such, especially oat-based products, oats being a source of soluble fibre, which is currently believed to be especially beneficial.

The desire for increased retention of vitamins and minerals has also fuelled the development of a number of initiatives. There has been much effort put into reducing the degree of processing involved in heat processed foods in order to reduce the amount of vitamin loss.

Environmental issues

The growth in the organic food market has been rapid over recent months and the number of products developed has been large (CIR, 1999). In 1999 a total of 567 new organic products were launched, compared with only 34 in 1997, and the trend is continuing. Although in the wake of intensification in agriculture, organic farming has continued to be practised, it has only recently re-emerged in a significant way. Initially, when organic produce came onto the supermarket shelves, the major driver appeared to be the consumer's desire to subscribe to what was perceived as the most environmentally friendly method of agriculture. But, as a niche market, the "barrier" for many consumers was price. Since that time, and several food scares later, the motivation now seems much more one of concern for safety and growing distrust of modern agricultural practice. Some consumers believe that organic produce is healthier, more nutritious and better tasting, although there has been little scientific research to investigate this.

The sector growth has also been aided by the entry of major food producers and retailers into the "organic" marketplace; it is therefore becoming more significant with established supply chains and a sustained demand. This has served to increase organic availability and erode the price differences between organic and conventionally grown materials. The fulfillment of this demand is largely through importation, especially into Northern European countries (e.g. the UK imports around 60-70% and Germany approximately 50% of organic foods on retail sale). At a European level, the market was valued at US$4.5 billion in 1997 and this is predicted to increase by around 20% per annum (CIR, 1999).

Organic farming is not the only environmental issue that influences consumer choice. The pressure to protect dolphins and other sea mammals in the commercial fishing of tuna has led to an international adoption of dolphin-friendly fishing techniques and associated logos. In the field of packaging, recycling, ozone-friendly, and sustainable forestry initiatives are now widely used as marketing tools.

Convenience issues

Until the 1960s, the convenience food market was limited to canned foods and dried foods. The rise in the ready meals market, and that of the extended snack, has been recent. Frozen ready meals started to appear in late 1960s and 1970s; the market quickly expanded and quality improved in line with consumer demands; these demands for improved quality then fuelled the rapid rise in chilled food product development. Recently, some of the major retailers have introduced ready meals claimed to be of restaurant quality - so you can have 'an evening out at home'! Also, now becoming available are meals to prepare or 'finish off' in the home, but with everything provided in the packet. The idea again is to provide improved final product quality with built-in convenience.

Closely related to the concept of convenience is that of shelf-life - the length of time the food will last for. Quality issues mean that the addition of preservatives and other additives, and more extensive processing are less preferred methods for extending shelf-life. Modified atmosphere packaging and other techniques have been extensively investigated to prolong the time that a food will remain in prime condition. There are now many products available that have extended shelf-lives through techniques that do not otherwise significantly affect the quality of the product.

6. QUALITY ASSURANCE AND QUALITY CONTROL

As has been discussed in the previous sections, there are many safety and quality issues that the food industry has to address. As well as the many individual types of hazard or quality issue, and processing and preservation mechanisms to control them, much time and thought needs to be put into overall management and design issues, in order that the individual operations can work successfully. Quality assurance is the philosophy of setting up management controls to assure that the end product is of the desired quality. Quality control is the testing of the end product to ensure that it meets stipulated requirements.

Safety is the primary concern of the food industry. The first priority is the well-being of the consumer and all other issues stem from the need to prevent unsafe food from being consumed. The best way of doing this is to prevent it from being made and distributed in the first place. The same is true of a product, which although not harmful, is not what the customer would expect (e.g. a vegetarian ready meal containing meat). The cost of a product recall if a problem is identified after release can in itself be very high, and there is the likelihood of prosecution if a problem is not recognised in time, especially if consumer health suffers as a result. The knock-on effect of loss of consumer confidence and subsequent loss of sales can be far more expensive. Some consumers are still influenced by an outbreak of typhoid in Aberdeen in 1964 which derived from post-process contamination of catering packs of corned beef and avoid purchasing such products. There have been recent cases of companies going out of business as a result of a safety issue. The company responsible for the hazelnut yoghurt botulism poisoning incident in the UK in 1992 were fined a relatively small amount, but were put out of business as a result of the incident.

In some cases, it is not just the particular company, but the whole of a product sector that may be affected. In the modern world of blazing tabloid headlines, a relatively minor issue can be blown out of proportion and the whole of a food sector or even the entire industry may suffer. Therefore, all responsible food companies make every effort to 'get it right every time'.

In addition to general company management philosophies, as exemplified in the BS ISO 9000 series of quality management standards (Rose, 2000), there are specific regimes and protocols that are now standard industry practice and apply specifically to food and drink production. These require constant surveillance, both internally and through audits by independent third parties.

6.1 The philosophy of HACCP

The internationally recognised philosophy used to help ensure the production of safe foods is HACCP (Hazard Analysis and Critical Control Points). EU hygiene directives state that food production processes must be evaluated and controlled using systems based on hazard analysis. In the UK, HACCP-based systems are now being introduced in all parts of the food supply chain, from agricultural production to retail sale. There have been several books published about the philosophy of HACCP (see Leaper, 1997; and Mortimore and Wallace, 1998 for examples). One of the main advantages of HACCP (which was originally devised by the Pillsbury Company to ensure the safety of food on board manned space missions) is that the need for end-product testing is reduced to a minimum and is replaced by systematic and targetted preventative measures. For example, it is not necessary to do exhaustive microbiological analysis of product if the HACCP study indicates that microbiological levels are being properly controlled (although some such testing may be required, together with specific measures for monitoring the effectiveness of the controls). This frees resources to enable targeted testing which can be more informative.

Although HACCP has historically been used for assuring product safety, it is now being used alongside other quality management systems to help assure product quality, authenticity and consistency of output. It is important that the objective and scope of each system is clearly defined. For example, in general terms, the philosophy of HACCP is to identify and control the specific hazards that could compromise product safety whereas the philosophy of ISO 9000 is to achieve consistent performance. Clearly HACCP can help in defining what performance is required, while ISO 9000 can help with the implementation of standardised procedures to achieve what is required. On a practical level, the acceptance of the culture of quality management through the one system will make implementation of the other much easier than it might otherwise be, as the relevant personnel will be familiar with the requirements for control of documentation and quality records, emphasis on compliance with systems and standards, the role of auditing and so on.

The HACCP approach is based on seven internationally recognized simple principles:

1. Conduct a hazard analysis: prepare a flow diagram of the steps in the process; identify and list the hazards associated with the process and specify how they are going to be controlled.

2. Determine the critical control points (CCPs), i.e. those stages at which hazard control is essential for the production of a safe end-product.

3. Establish critical limits for each hazard at each CCP, i.e. the levels for each individual hazard that must not be exceeded if a safe product is going to be achieved. This may, for example, be a maximum storage temperature, or a maximum limit for a particular micro-organism, or a minimum salt level in the product.

4. Set up a system to monitor control of each CCP by scheduled testing and observations, to ensure that the hazard remains within critical limits.

5. Establish what corrective action needs to be taken if monitoring indicates that a particular CCP is not under control *or is moving out of control,* i.e. is going beyond critical limits - this means stopping something going wrong before it happens, if at all possible.

6. Set up procedures to make sure that the overall HACCP plan is working as desired; this may include some end-product testing and a regular review of the system.

7. Establish thorough documentation of the system, process and procedures, and of all measurements taken relating to the monitoring of the process.

Software programmes (such as CCFRA's *HACCP Documentation Software and safefood Process Design System*) are available to help companies formulate and record their HACCP systems.

Application of HACCP - example

Although production of fresh produce (e.g. lettuce) and a manufactured product (e.g. a cook-chilled ready meal) are completely different operations, the philosophy and practice of HACCP can be applied equally to both. In both cases, HACCP can be used to identify hazards, determine control points, establish limits, monitor against these, plan remedial actions in the event of a problem and keep records. HACCP is a generic framework that can fit all processes. The benefit for users is that they can think through their own processes in a structured, standardised way. This small extract from each of two example HACCP analyses illustrates this.

	Typical Hazards	Control Measures
Lettuce	Contamination with unacceptable levels of pesticide during cultivation	Correct training of operators Follow approved codes of practice
	Contamination with pathogens during irrigation	Use approved source of irrigation water
	Contamination with glass from equipment during harvesting	Draw up glass policy and breakage policy
	Growth of vegetative pathogens during transport due to incorrect temperature	Control temperature of vehicle
Chicken ready meal with garlic puree	Growth of pathogens during chill storage of chicken	Control chiller temperature within specified limits
	Breakage of metal blade during slicing of chicken	Visual inspection of knives before use
	Presence of pathogens in purchased garlic ambient stable puree	Supplier quality assurance programme
	Growth of pathogens during stuffing operation	Control room temperature Control time of stuffing operation prior to cooking

References:

Knight, C. and Stanley, R.P. (1999) Assured crop production: HACCP in agriculture and horticulture. CCFRA Guideline No. 10. Supplement 1: Lettuce case studies.

Leaper, S. (1997) HACCP: A prtactical guide. Technical Manual No. 38, 2nd Edition. CCFRA.

6.2 Hygienic design and good manufacturing practice

Paraphrasing the EU food hygiene directive, food hygiene involves all measures (to prevent contamination) to ensure the safety and wholesomeness of foods and covers all stages after primary production, including offering for sale or supply to the consumer. Although this section focusses on microbial issues, chemical and physical contamination must also be addressed. Micro-organisms are ubiquitous and will be initially present on all raw materials, as well as in production areas and on processing equipment. The most important feature of micro-organisms is that, given suitable conditions, they will multiply rapidly. The important aim of the food industry is to prevent unacceptable levels of unwanted micro-organisms, especially pathogens, from remaining in the product at the end of processing. This is especially important in food which is not going to be cooked before eating.

Food can be processed in several ways to control or eliminate viable micro-organisms, as has been described in Chapter 2. However, there are many additional controls that the food processor has to take to ensure a safe product. All processing needs to be carried out in a hygienic manner, i.e. in a way that reduces to a minimum the chances of microbial cross-contamination from the environment in general to the food. There are many guidelines and codes of practice detailing how best to do this. General guidelines such as those produced by the Institute of Food Science and Technology (IFST, 1998) are widely used and accepted in the industry. There are also guidelines for specific food types, and for specific processes, such as that for heat-preserved foods produced by the Department of Health (DoH, 1994). In simple terms, these guidelines detail all the relevant issues that have to be addressed in the manufacture of safe and palatable foods.

For the food manufacturer, the first prerequisite is to site the premises in a suitable location - e.g. one that is not liable to flooding and is not in an area with significant air pollution that might taint or contaminate product.

Depending on the type of process being carried out, the factory plant has to be properly laid out. This means designing floors, walls, ceilings and services in the correct way (see Timperley and Timperley, 1993; and Timperley, 1994) and preventing entry of pests such as rodents, insects and birds. The equipment itself has to be correctly designed, for example, avoiding 'dead-ends', badly fitting valves and poor connections which might harbour residual product and result in microbial growth (see Timperley, 1997 for guidance on liquid-handling equipment). This is very much a partnership between the food producer and the equipment manufacturer to ensure that the equipment can be easily cleaned and is generally fit for the purpose. As well as the design of the equipment, it has to be correctly laid out in the factory, for example, to prevent raw materials from entering finished product areas, which might result in microbial cross-contamination.

Having set up the production area, it has to be properly maintained. This means efficient cleaning and disinfection of all process areas and equipment, using the right materials and the appropriate tools. Studies have shown that some methods of cleaning (e.g. with high pressure hoses) can disperse contamination over a wide area, effectively negating the cleaning procedure and their use must be carefully controlled. Air movements also have to be considered, especially in 'high-care' areas such as chilled food production areas (see Brown, 1996). Micro-organisms are present in the air and it is important that there is no significant deposition onto processed food. This means, among other things, preventing condensation from occurring in the processing area and, in some instances, directing air flow away from the processing line.

Personnel hygiene is also very important. Hand washing procedures have to be formulated and adhered to; beards and hair have to be netted; and there has to be strict controls to prevent workers with gastrointestinal illnesses from entering the production area.

All of these hygiene and manufacturing considerations have to be addressed if the individual processing and preservation operations are to be effective.

Restricted airflow caused spoilage

Genetic fingerprinting techniques have illustrated how important factory lay-out can be, and how changes can have unforeseen effects that compromise product safety or quality. A food company was having problems with product spoilage, and conventional microbiological analysis had implicated *Pseudomonas*, but could not pinpoint the source. Using a fingerprinting technique, the strain isolated from the spoiled food was matched exactly with that found in condensation on processing equipment. Further investigation revealed that the condensation had arisen because of changes in plant layout which had reduced airflow. Having identified the cause, remedial action was taken to increase airflow in relevant areas and so eliminate the problem.

Good handwashing technique

A correct hand-washing regime is very important in maintaining adequate hygiene in food processing factories, especially in 'high-care' areas, such as chilled food production facilities. Hands should be wetted and soap used from a dispenser. All parts of the hands and wrists should be rubbed, as shown below, with each step consisting of 5 strokes forward and backward.

1. Palm to palm

2. Right palm over left dorsum and then left over right

3. Palm to palm - fingers interlaced

4. Backs of fingers to opposing palms, interlocking the fingers

5. Rotational rubbing of right thumb and vice versa

6. Rotational rubbing backwards and forwards - use the fingers of right hand on left palm and vice versa

References:

Ayliffe, G.A.J., Babb, J.R. and Quaraishi, A.H. (1978) A test of hygienic hand disinfection. Journal of Clinical Pathology, **31**: 923-928.

Taylor, J.H. and Holah, J.T. (2000) Hand hygiene in the food industry: a review. CCFRA Review No. 18.

7. FOOD LEGISLATION AND REGULATION

There seems, at times, to be a widespread perception that food and food production is subject to relatively little legislative control. In reality, however, food and food production is subject to detailed and extensive regulatory controls which impinge on everything from crop and livestock husbandry to the labelling and sale of the final product. Legislation varies considerably from country to country or between economic regions (e.g. EU vs USA), but the purposes of controls are usually much the same:

- To help assure food safety and minimise foodborne illness
- To promote fair trade and prevent consumers and the industry from being misled or defrauded
- To provide consumers with useful and accurate information on ingredients, additives, nutritional content and 'ethical' issues

Because of the variation from region to region, it is not possible in a book of this size to describe the legislation in great detail as it applies throughout the world. The purpose of this chapter is to illustrate, by providing a description of some of the controls in place in the UK, how legislation can be used to control certain practices and how legislation works alongside voluntary controls. This section, therefore, provides outline examples of some UK controls (many of which originate from the EU), but it is not a definitive reference guide.

7.1 The legislation - some examples

The controls on all matters relating to the food we eat are a mixture of statutory legislation, industry self-regulation, and guidelines and codes of practice. Legislation is passed by the European Union and by the UK government and devolved assemblies. Industry codes and guidelines are written by trade associations, industrial bodies, and international organisations such as the Codex

Alimentarius Commission, and the day-to-day enforcement of the system is managed by the local authorities. However, the interpretation of the law is ultimately a matter for the Courts.

Food legislation in the UK may be traced back to the Middle Ages and the development of modern food law has been an evolutionary process. As government perceived a need for further measures of control on food, Acts of Parliament were passed and in later years a multiplicity of secondary legislation (i.e. regulations/ orders) were made to control the manufacture, safety, packaging and description of food. However, since the accession of the UK to the Treaty of Rome in 1972 there has been a gradual harmonisation of UK food law with legislation of the European Union (EU). Today most UK food law derives from EU Directives and Regulations, although it is still implemented under the umbrella of the UK legal system.

An EU Directive is a requirement by the Council of the EU that all Member States shall adopt national measures to meet its objectives, within a specified time limit. In contrast, EU Regulations are directly binding on the Member States to whom they are addressed and do not require national enactment. In addition, the EU often issues Recommendations and Opinions, which are just that and are given by the European Parliament on proposed legislation or similarly by various committees of the EU Institutions.

Although most current UK food law derives from EU legislation, it is still incorporated into UK law under its traditional structure. Primary legislation is made via Acts of Parliament. These are usually very general in their nature and give Ministers certain powers to make more specific pieces of legislation. The drafting and passing of Acts of Parliament are lengthy processes, and their rapid amendment is rarely possible. Therefore, the day-to-day enactment of legislation is carried out via Statutory Instruments, which are drafted under the auspices of a specific Act. These Statutory Instruments may then be amended, if required, without the need to rewrite and republish the entire legislation.

Guidelines and Codes of Practice written by trade associations and other non-governmental bodies are generally subservient to the Statutory legislation and are basically 'how-best-to-do-it' instructions. These are usually voluntary guidelines, compliance with which will help companies to comply with statutory legislation. However, there are some statutory government codes of practice including a series relating to matters in the Food Safety Act.

7.1.1 Food Safety Act 1990

The Food Safety Act 1990 is the major piece of legislation covering matters relating to food production and retailing. It was not directly derived from EU legislation, but is compatible with all existing EU requirements, and allows more specific EU controls to be given effect in UK law. There are other Acts which impinge on the food chain, which will be mentioned later on.

There are numerous basic provisions made under the Food Safety Act. Some of the most important points are described below. In many cases, a food might be regarded as failing to meet more than one statutory requirement, and there is sometimes uncertainty over which section of the Act to make a prosecution.

Food safety requirement

The concept of a 'food safety requirement', in simple terms, states that it is unlawful to sell, or even offer for sale, any food which is unfit for human consumption, or has been rendered harmful to health by the addition of an ingredient or the removal of any constituent or subjecting it to any process or treatment, or has been so contaminated that it would be unreasonable to expect it to be used for food consumption in that state. This might cover such diverse issues as a microbial pathogen, foreign body contamination or high levels of potentially toxic heavy metals or pesticides.

Substance, nature or quality

It is illegal to sell food which is inferior or substantially different to what it is purported to be. Thus, it is not permissible to sell or advertise ordinary long grain rice as Basmati rice, or pork and beef sausages as all-pork sausages, or non-wholemeal brown bread as wholemeal. It is also illegal to sell food that contains an ingredient or other component which is not compatible with the food demanded, or food that falls below the quality that an ordinary purchaser would expect to receive. There are still some compositional standards in UK law (e.g. for jams and some meat products - see pp.99-100), but in other cases courts of law might be invited to set a minimum purchaser's standard. In many cases there will be industry guidelines for composition (e.g. for the amount and variety of fruit in a fruit cocktail or fruit salad) and these could be used in the decision making.

Misleading descriptions

It is an offence to make a false or misleading claim on a food label or in an advertisement for a food. The 'label' does not have to be attached to the food (e.g. it could be as a general label on a delicatessen counter which covers all of the products therein). A label is false if there is a factual mis-statement; it could be misleading if it implies by inclusion or omission something which is untrue. The latter might include describing something as British, when the major ingredient derives from abroad and was merely packaged in Britain.

Misleading presentation

This covers such things as the shape, appearance and packaging of a food, the way in which it is arranged for sale and the setting in which it is displayed. Examples of misleading presentation would include displaying fatty mince under red lighting to give a misleading impression of its fat content; displaying analogue dairy products alongside genuine milk-based products; and packing artificially flavoured fruit products in fruit-shaped containers. It should be noted that truthful labelling of products may, in some cases, mitigate or negate a misleading presentation offence.

Food Safety Act due diligence defence

Probably one of the most often quoted pieces from the Food Safety Act is that relating to 'Due Diligence'. The Act states that if it is proven that someone failed to meet the requirements of the Act, "... it shall be a defence for the person charged to prove that they took all reasonable precautions and exercised all due diligence to avoid the commission of the offence by himself or a person under his control".

This means a person or company cannot be found guilty if they did everything reasonably possible to avoid the offence being committed. What 'all reasonable precautions' are will vary from case to case, and will be influenced by the facilities available to the person concerned. For example, a large manufacturing concern might be expected to put in place more exacting and expensive monitoring equipment than a small manufacturer.

7.1.2 Individual legislation under the umbrella of the Food Safety Act

There are approximately 300 individual pieces of legislation in the UK relating to food and food production, the majority of which come under the umbrella of the Food Safety Act. These pieces of secondary legislation are often called Statutory Instruments, and they provide the day-to-day detail of what is and is not allowed to be done to the food and drink that we consume. Examples of some of the issues covered by those Statutory Instruments are outlined below.

Additives

Derived from EU Directives there is legislation covering the addition of colours, artificial sweeteners, and 'miscellaneous' additives to food intended for sale to the public. The 'miscellaneous' category includes such things as preservatives, antioxidants, emulsifiers, stabilisers, acidulants, humectants, anti-caking agents, anti-foaming agents, flavour enhancers and thickeners amongst others. (For details on the role of these additives in food, see Chapter 1.3). The legislation includes positive lists of what is permitted to be added to foods, and in many cases there are restrictions on the amount that can be added and the foods to which they can be added. In addition, there is legislation covering the presence of associated materials such as extraction solvents, flavourings and mineral hydrocarbons.

Compositional standards

Compositional standards were originally introduced to maintain quality in specific types of formulated and processed foods and to prevent false descriptions being applied to food types. For example, there was legislation covering exactly what constituted (in terms of fat content and other characteristics) double, clotted and other cream categories. There has been a tendency in recent years to move away

What is jam?

The specific requirements for calling a product 'jam' are a good example of the requirements of compositional standards. The following prescriptions are taken from the 'Jam and similar products regulations, 1981 (SI 1063)':

Jam A mixture, brought to a suitable gelled consistency, of sweetening agents and fruit pulp or fruit puree, or both, such that:

a) the quantity of the fruit pulp and fruit puree used for every kilogram of finished product is not less than:

in the case of passion fruit, 60 grams,
in the case of cashew apples, 160 grams
in the case of ginger, 150 grams
in the case of blackcurrants, rosehips or quinces, 250 grams, and
in the case of any other fruit, 350 grams; and

b) the soluble solids content of the finished product, determined by refractometer at 20°C, is not less than 60%

In addition to these prescriptions, there are also specific requirements for extra jam, jelly, extra jelly, chestnut puree, reduced sugar jam, reduced sugar jelly, marmalade, reduced sugar marmalade, UK standard jelly, X curd, Y flavour curd and mincemeat.

from prescriptive definitions of the composition of certain food types, especially those for very specific products which are better controlled by voluntary codes of practice and custom and usage. Examples of this are butter and most aspects of the composition of margarine, as well as the categories of cream. In their place, requirements have been introduced for more informative labelling of specific foods. Thus food manufacturers have had the freedom to develop and market a much wider range of food products to meet ever changing consumer requirements, and consumers have been able to make an informed purchasing decision from the much broader range of foods now on offer.

However, several categories of food still have clearly defined requirements. These include: bread and flour (e.g. minimum requirements to label bread as wholemeal); cocoa and chocolate products (e.g. minimum levels of chocolate solids); coffee products; fruit juices and nectars; honey; jams and similar products (e.g. definitions of extra jam and marmalade and reduced-sugar products); meat products (e.g. definitions of meat content in burgers, sausages and pies); and spreadable fats (vitamin content in margarine). In-depth details of the composition and labelling of many specific food products have been compiled in CCFRA's multi-media CD-RoM, *Enlabel*, and more general information is given in Food Law Notes (CCFRA, 1999).

Contaminants

In addition to the general Contaminants in Food Regulations that control nitrates and mycotoxins in certain commodities, the over-riding requirements of the Food Safety Act state that food must not be so contaminated as to be unfit for human consumption. There are specific pieces of legislation covering various contaminants in food. These include arsenic, chloroform, erucic acid, lead, tetrachloroethylene, tin and tryptophan and detail the maximum level of each allowed in specific foodstuffs.

Hygiene

Production of food in a hygienic and safe manner is an over-riding concern of the food industry and there has been much EU-wide legislation to control such matters. In the UK this has been enacted via the Food Safety (General Food Hygiene)

Regulations, and individual regulations covering dairy products, egg products, fish and shellfish products, and fresh meat, minced meat, poultry meat, wild game and meat products. These latter regulations are extensive in their requirements and in many cases include prescribed maximum levels for specified micro-organisms (e.g. *Listeria, Salmonella* and coliforms).

Labelling

Food must be so labelled that the customer can make an informed choice as to whether to buy it or not. The Food Labelling Regulations 1996, as amended, prescribe many requirements for a food label. Table 11 lists some examples. Amongst these are the need for a clear product name, and a supplementary description if a 'fancy' name is used (e.g. Cadbury's 'MiniRolls', where 'MiniRoll' is the fancy name, have the supplementary description: "Individual milk chocolate covered Swiss roll with vanilla flavour filling"). Also required in most cases are a weight or volume marking, a list of ingredients, and in some cases a table describing

Table 11 - Some basic considerations when labelling a product

1. Is the product name clear?
2. Is a supplementary name required?
3. Does the product require a 'use by' or 'best before' date?
4. Is nutrition information required?
5. If so, is it correctly presented and accurate?
6. Is the weight/volume declaration correct and clear?
7. Is the product packaged to the average weight system?
8. Is quantitative ingredient declaration required?
9. Are product name, weight/volume and use by/best before in the same field of view?
10. Are any pictures on the label clear and not misleading?
11. Is there a manufacturer's, packer's or seller's name and address?
12. Is the overall label clear and fair?
13. Does it, in fact, look very like another company's label for a similar product?

the nutritional content of the food. Recently, a need to quantify the amount of certain ingredients in some products has been introduced - the so-called QUID (Quantitative Ingredient Declaration) regulations.

In addition to factual information, the label must not give misleading information, (e.g. through misleading pictorial representation) or be so badly presented as to make the true nature of the information difficult to ascertain.

There are many exemptions to some or all or the Food Labelling Regulations for certain types and forms of food, and many additional specific requirements for other food types. To help unravel the complexities and subtleties of the legislation, CCFRA produced a multi-media CD-RoM, called *Enlabel*, which uses text, graphics, video and voice-over to guide the user through this potentially complex subject.

Options for date marking of product

The form in which date marks are given depends upon the type and shelf-life of a food. Short shelf-life products that are likely to become hazardous due to the action of micro-organisms must be marked "use by", together with any storage conditions that should be observed. The date specified should include the date and month - for example, 'Use by 30 June'.

Other products that require a date mark are labelled "Best before".

Products with a shelf-life of less than 3 months are marked "Best before" with the date, month and, optionally, year. In this case, both 'Best before 30 June' and 'Best before 30 June 2000' would be acceptable.

Products with a shelf-life of between 3 and 18 months can be marked "Best before end" with the date alternatively specified as just month and year - for example 'Best before end June 2000' would be equally as acceptable as 'Best before 30 June 2000.

Products with a shelf-life of more than 18 months can be marked "Best before end" with the further alternative of just specifying the year. In this case the date could be given as just 'Best before end 2000', or as 'Best before end December 2000' or as 'Best before 31 December 2000'.

Packaging and food contact issues

Packaging is an important part of the product. It protects food from contamination, helps prevent deterioration of the product and enables product information to be given to the potential purchaser. By virtue of their function, many packaging materials come into contact with the food.

Any food packaging or other material that comes into direct contact with food must not adversely affect the food to a material degree. There are two specific pieces of legislation in the UK covering these matters: the general 'Materials and Articles in Contact with Food Regulations 1987', as amended, and the 'Plastic Materials in Contact with Food Regulations 1998'. The latter gives a list of monomers that are permitted to be used in the production of food contact plastics, and there are both overall and specific limits for the amount of transfer of these and other materials into the food.

Other Acts of relevance to food production

In addition to the Food Safety Act there are several other Acts which are not specific to food, but which nevertheless are of significance to the food industry (see Table 12). One good is example of this is the Weights and Measures Act. The Statutory Instruments made under this act cover, amongst other issues, all matters relating to the quantities (volume or weight) that a food can be packaged in. They also cover the marking of food quantities in metric and/or imperial units, and the need, in certain situations, to label the price per unit weight, volume or number (e.g. 8.4p per 100ml on a litre of squash priced at 84p).

Matters related to the 'Average Weight System' (**e** mark) are also covered. Under this legislation, packers of a food product are allowed a tolerance on the actual weight or volume in a package, provided that the average weight of a package in a bulk lot of packages does not fall below that stated on the package. For example, for product packs labelled 500g **e** (the nominal quantity), the average quantity in these must be at least 500g, but individual packs may be up to 15g below this. Also up to 2.5% of packs may be up to 30g below, but none shall be more than 30g below. This allows for the inevitable variation in pack-to-pack filling but also ensures that the consumer gets a fair deal.

Some foods have to be packaged in prescribed quantities (see Table 13). For example, flour must be sold in quantities of 125g, 250g, 500g or multiples of 500g (cornflour can also be sold in 375g and 750g quantities), whereas quantities of bread over 300g must be sold in multiples of 400g.

Table 12 - Examples of other Acts of relevance to the food industry

Food and Environment Protection Act

This Act covers, amongst other things, matters relating to pesticide control and usage, including permitted maximum residue levels on foods.

Health and Safety at Work etc Act

All food businesses have to take into account the well-being of their employees during their operation. See Brown (2000) for an example concerning the temperature of the working environment during chilled food production.

Water Industry Act and Water Resources Act

There is legislation made under these Acts on the quality of water supplies, so assuring the quality of incoming water at a food production plant. See also Dawson (1998).

Trade Descriptions Act

The Trade Descriptions Act has many provisions that are analogous to the 'substance, nature and quality' provisions of the Food Safety Act, and covers quantity, method of manufacture, composition and fitness for purpose among others.

Consumer Protection Act

One specific piece of legislation made under the Consumer Protection Act is that covering price marking, which requires that the selling price of goods for retail sale be indicated. It also requires the unit price for many products to be indicated (e.g. price per 100g, or per litre).

Agriculture Act

Various pieces of legislation pertaining to feedingstuffs are made under the Agriculture Act. One such, The Feedingstuffs Regulations, covers the composition of feedingstuffs, their labelling and allowed additives.

Animal Health Act and Welfare of Animals at Slaughter Act

Covers, amongst other issues, the wellbeing of animals reared for food production.

Table 13 - Examples of foods that must be sold in prescribed quantities

Food	Prescribed quantity*
Biscuits	100g and multiples thereof, 125g, 150g, 250g
Bread	400g and multiples thereof
Breakfast cereals (except biscuit forms)	125g, 250g, 375g, 500g, 750g, 1kg, 1.5kg and multiples of 1kg
Flour	125g, 250g, 500g and multiples thereof
Jam/marmalade	57g, 113g, 227g, 340g, 454g and multiples thereof, 680g

* In all cases, there are upper and/or lower limits, outside of which these restrictions do not apply

7.2 Enforcement of legislation

In the UK, the day-to-day enforcement of food legislation is carried out by local authorities' enforcement officers. In the Shire counties and Unitary Authorities of England and Wales, the trading standards officers deal with the labelling of food, its composition and most cases of chemical contamination, as well as weights and measures. Environmental health officers deal with hygiene matters, cases of extraneous matter in food, microbiological contamination of foods and food which, for any reason including chemical contamination, is unfit for human consumption. They also oversee general health and safety matters. In London and the metropolitan areas, these functions are often combined within a single department. In Scotland, all food law enforcement is carried out by the environmental health departments of district and island councils. The UK Government also plays a role. In addition to becoming involved in emergencies, the newly created Food Standards Agency has an important role in overseeing local authority food law enforcement activities (FSA, 2000). This includes, for example, setting and monitoring performance standards and auditing local authorities' food law enforcement activities for their effectiveness and consistency. The activities of individual trading standards and environmental health departments in matters relating to food is coordinated by LACOTS (The Local Authorities Coordinating Body on Food and Trading Standards).

7.2.1 The principle of the Home Authority

The Home Authority Principle has been developed by LACOTS as an aid to good enforcement practice: practices which protect the consumer, and encourage fair trading, consistency and common sense. The Home Authority is the authority (Local Council) where the decision-making base (i.e. the headquarters) of a business is located and it is this authority that would coordinate any investigations or prosecutions into possible breaches of the law. As an example, if a retailer with several branches around the country were to have a problem with a labelling issue, the Home Authority might coordinate an investigation and issue the company with details of its findings. This might be that the Authority found the labelling to be incorrect and that amendments were required within a certain time-scale. This decision would be relayed to all authorities in which the retailer had outlets, and these authorities could then monitor the company's response. Clearly, this is a much more streamlined system than having individual authorities each investigating the same problem and, possibly, separately issuing different guidance notes.

The Home Authority Principle commands the support of local authorities, central government, trade and industry associations, and consumer and professional regulatory bodies and aims to:

- encourage authorities to place special emphasis on goods and services originating within their area
- provide businesses with a home authority source of guidance and advice
- support efficient liaison between local authorities
- provide a system for the resolution of problems and disputes.

7.2.2 Enforcement tools

In matters relating to food, the Food Safety Act contains extensive powers to enable food authorities to deal effectively with problems. A series of about 20 Statutory Codes of Practice have been issued under the auspices of the Act to guide enforcement officers in their duties both towards the provisions of the Act itself and many of the individual pieces of legislation made under the Act.

Ministry guidelines and FAC recommendations

In addition to specific pieces of legislation, Ministers may find it appropriate to issue non-statutory guidelines, sometimes based on recommendations by the Food Advisory Committee (FAC). The FAC is one of several committees set up to advise the UK government on particular aspects of food policy. The Committee is concerned with the labelling, composition and chemical safety of food. Before the introduction of new legislation, the FAC may be asked to conduct a review and to make recommendations, which generally are then published as a report. In the absence of specific legislation, FAC recommendations or views and those of its predecessors, the Food Standards Committee (FSC) and the Food Additives and Contaminants Committee (FACC) may be used by enforcement authorities to assess whether or not there is a case for prosecution. This is more often the case where the recommendations have gained the acceptance of Ministers.

Case law

Case law arises from past decisions of the higher courts and these decisions are often quoted by both the prosecution and the defence to lend support to their arguments. Such decisions may be used as guidance by local authorities in deciding whether or not to prosecute as well as by the courts when considering the evidence presented during hearings.

LACOTS opinions

In addition to formal legislation, opinions and guidelines from government departments, and published voluntary codes of practice (see below), LACOTS also issues opinions on interpretation of various practices in the industry. These are a very good indication to industry of what issues the enforcement authorities are likely to be concerned about and where warnings or prosecutions may be issued if companies step outside a set of guidelines. As with other opinions, however, they are not legally binding as only the courts can interpret the law with authority.

7.3 Self-regulation and codes of practice

It would be impossible for government to issue legislation that adequately covered all the minutiae associated with food production. Specific areas where industry self-regulation is of major importance in the production and retailing of food are: hygienic design of equipment and processing regimes; the naming and description of food products; and specifications for certain aspects of product quality. In these areas, codes of practice, standards and guidance notes written by and for the industry are of great importance.

The guidelines used by industry are issued and authorised by numerous organisations, including professional bodies such as the Institute of Food Science and Technology or the Codex Alimentarius Commission (a body responsible to the World Health Organisation and the Food and Agriculture Organisation), the Research Associations, or one of the many sector-specific Trade Associations. However, the technical content is usually provided by technical experts from within the industry. A selection of these, giving a flavour of the topics covered, are listed in Table 14.

Table 14 - Examples of codes of practice

Food and drink. Good manufacturing practice. A guide to its responsible management - 4th Edition (Institute of Food Science and Technology, 1998)

Guidelines for the safe production of heat preserved foods (Department of Health, 1994)

Guidelines for good hygienic practice in the manufacture of chilled foods (Chilled Foods Association, 1997)

Guidelines for small-scale fruit and vegetable processors (Food and Agriculture Organisation of the United Nations, 1997)

Food packaging hygiene standard (PIRA, 1995)

Guidelines for the facilities and equipment required for handling bivalve molluscs from harvesting through to distribution to retail outlets (Sea Fish Industry Authority, 1997)

Code of practice on bread weight checking for plant bakers (Federation of Bakers, 1995)

Code of practice for the hygienic manufacture of ice cream (Ice Cream Alliance and Ice Cream Federation, 1995)

A code of practice relating to the transportation of wines, spirits and concentrated grape must in bulk (The Wine and Spirit Association of Great Britain and Northern Ireland, 1994)

Guidelines for good hygienic practice in the manufacture of dairy-based products (Dairy Industry Federation, 1995)

Code of practice for the packaging of consumer goods (Packaging Standards Council, 1994)

Code of practice for the soluble coffee industry in the UK (British Soluble Coffee Manufacturers' Association, 1995 - ISBN 0952586703)

The control of pesticides: a code of practice (Fresh Produce Consortium, 1997)

Guidelines for the exercise of due diligence in the manufacture, storage and distribution of meat and other food products (British Meat Manufacturers' Association, 1996)

Code of practice and minimum standards for sandwich bars and retailers of packaged sandwiches (British Sandwich Association, 1997)

The Portman Group code of practice on the naming, packaging and merchandising of alcoholic drinks (The Portman Group, 1997)

Fresh-cut produce handling guidelines (International Fresh-Cut Produce Association, 1999)

Code of practice for the good hygienic manufacture and distribution of food additives and ingredients (Food Additives and Ingredients Association, 1998)

8. REFERENCES AND FURTHER READING

Adams, J. (1999) Cars, cholera, cows and contaminated land - virtual risk and the management of uncertainty. In: *What Risk? Science, Politics and Public Health*, Bate, R. (ed.), Butterworth Heinemann, pp. 295-304, ISBN 0 7506 4228 9.

Air Products (1995) The Freshline guide to modified atmosphere packaging.

Air Products (1999) Modified Atmosphere Packaging Gas Selector. CD-RoM.

Anon (1990) Exploited plants: collected papers from Biologist. Institute of Biology, London.

Anon (1998) Food and Drink Statistics. EuroPA and Associates.

Anon (2000) *Salmonella* infections in humans, England and Wales: 1981 to 1999. CDR Weekly, **10** (6): 50

Arribas-Jimeno, S. (1982) The Spanish toxic syndrome. Trends in Analytical Chemistry 1 (14): iv-vi.

Bedford, L.V. (1986) Leguminous crops and their pulse products: a guide for the processing industries. Technical Bulletin 59. Campden & Chorleywood Food Research Association.

Bell, C. and Kyriakides, A.(1998a) *Escherichia coli*. A practical approach to the organism and its control in foods. Blackie Academic and Professional

Bell, C. and Kyriakides, A.(1998b) *Listeria*. A practical approach to the organism and its control in foods. Blackie Academic and Professional.

Bell, C. and Kyriakides, A.(2000) *Clostridium botulinum*. A practical approach to the organism and its control in foods. Blackwell Science.

Betts, G.D. (1996) A code of practice for the manufacture of vacuum and modified atmosphere packaged chilled foods with particular regard to the risks of botulism. Guideline No. 11. Campden & Chorleywood Food Research Association.

Birch, G.G., Blakeborough, N. and Parker, K.J. (1981) Enzymes and Food Processing (Applied Science Publishers).

Brown, K.L. (1996) Guidelines on air quality standards for the food industry. Campden & Chorleywood Food Research Association.

Brown, K.L. (2000) Guidance on achieving reasonable working temperatures and conditions during production of chilled foods. Guideline No. 26. Campden & Chorleywood Food Research Association.

Campbell, A.J. (1991) The shelf-stable packaging of thermally processed foods in semi-rigid plastic barrier containers. Technical Manual No. 31. Campden & Chorleywood Food Research Association.

Campbell, A.J. (1995) Guidelines for the prevention and control of foreign bodies in food. Guideline No. 5. Campden & Chorleywood Food Research Association.

CCFRA (1999) Focus on Sandwiches. A compilation of data from the "NewFoods" CD-ROM and Product Intelligence Exhibition Service, July 1999.

CCFRA (1992) Retort operations: a video training programme.

CCFRA (1999) Food Law Notes. A reference manual on UK Food Law.

CIR (1999) The European Market for Organic Foods. Corporate Intelligence on Retailing, 48 Bedford Square, London, WC1B 3DP.

Craven, B.M. and Stewart, G.T. (1999) Public policy and public health: coping with potential medical disaster. In: What Risk? Science, Politics and Public Health, Butterworth Heinemann, pp. 222-241, ISBN 0 7506 4228 9.

Cunningham, A. (1991) Essential British history: key dates, facts and people summarised. Usborne. ISBN 0 7460 0658 6.

Cybulska, G.I. (2000) Waste management in the food industry: an overview. Key Topics in Food Science and Technology No. 2. CCFRA. ISBN: 0 905942 30 2.

Dawson, D. (1998) Water quality for the food industry: an introductory manual. Guideline No. 21. Campden & Chorleywood Food Research Association.

Dawson, D. (2000) Water quality for the food industry: part 2. Guideline 27. Campden & Chorleywood Food Research Association.

Day, B.P.F. (1992a) Guidelines for modified atmosphere packaging. Technical Manual No. 34. Campden & Chorleywood Food Research Association.

Day, B.P.F. (1992b) Chilled food packaging. In: Chilled Foods: A Comprehensive Guide (Eds. C. Dennis and M. Stringer). Ellis Horwood.

Day, B.P.F. (2000) Chilled food packaging. In: Chilled Foods: A Comprehensive Guide (Eds. M. Stringer and C. Dennis). Woodhead Publishing.

Department of Health (1991) Dietary reference values for food energy and nutrients for the United Kingdom. Report on Health and Social Subjects 41.

Department of Health (1994) Guidelines for the safe production of heat-preserved foods.

DETR (2000) Waste strategy 2000 for England and Wales.

Dixon, B. (1994) Enzymes make the world go around. Novo Nordisk, Denmark.

Duddington, C.L. (1969) Useful Plants. McGraw Hill.

Edwards, M.C. and Redpath, S.A. (1995) Identification of foreign bodies from food. Guideline No. 4. Campden & Chorleywood Food Research Association.

Edwards, M.C. and Fincher, C.H. (1999) The effect of food processing on foreign bodies: a case study on baking. Review No. 13. Campden & Chorleywood Food Research Association.

Emsley, J. (1994) The consumer's good chemical guide. A jargon-free guide to the chemicals of everyday life. W.H. Freeman. ISBN: 0 7167 4505 4.

Emsley, J. and Fell, P. (1999) Was it something you ate? Food Intolerance: What Causes It and How to Avoid It. Oxford University Press. ISBN 0 19 850443 8.

Farnsworth, M.W. (1978) Genetics. Harper and Row. ISBN 0 06 042003 1.

Fox, P.F. (1993) Cheese: chemistry, physics and microbiology. Volume 1. Chapman and Hall.

FSA (2000) Framework agreement on local authority food law enforcement. Food Standards Agency. www.foodstandards.gov.uk

Guy, R.C.E., Hodge, D.G. and Robb, J. (1993) An examination of the phenomena associated with cake staling. FMBRA Research Report No. 107.

Hall, M.N. and Jones, J.L. (1999) Public confidence in fertilizers and in food quality and safety. International Fertiliser Society. Proceeding No. 441.
ISBN 0 85310 076 4.

Harrison, M., Llewellyn-Davies, D. and Everitt, M. (2000) New products 1999 - 12 month review. CCFRA.

Heywood, V.H. and Chant, S.R. (Eds) (1982) Popular Encyclopaedia of Plants. Cambridge University Press. ISBN 0521 24611 3.

Holland, B., Welch, A.A., Unwin, I.D., Buss, D.H. Paul, A.A. and Southgate, D.A.T. (1991) McCance and Widdowson's Composition of Foods. 5th Edition. Royal Society of Chemistry and MAFF.

Hoseney, C. and Miller, R. (1998) Current understanding of staling of bread. American Institute of Baking Research Department Technical Bulletin **20** (6): 1-6.

IFST (1993) Listing of codes of practice applicable to foods. Institute of Food Science and Technology.

IFST (1998) Food and Drink. Good Manufacturing Practice. A guide to its responsible management. Institute of Food Science and Technology.

IFST (1999) Position paper on bovine spongiform encephalopathy (BSE) (June 1999 edition). Institute of Food Science and Technology, 5 Cambridge Court, 210 Shepherds Bush Road, London, W6 7NJ.

Ironside, J.W. (1999) nvCJD: exploring the limits of our understanding. Biologist **46** (4), 172-176.

Jones, J.L. (1996) Food biotechnology: current developments and the need for awareness. Nutrition and Food Science (6): 5-11.

Jones, J.L. (1999) The food, the fad and the technology. Biologist **46** (3): 144.

Jones, L. (1999) Molecular methods in food analysis: principles and examples. Key Topics in Food Science and Technology No. 1. CCFRA. ISBN: 0 905942 28 0.

Langseth, L. (1995) Oxidants, antioxidants and disease prevention. International Life Sciences Institute Monograph.

Kent, N.L. and Evers, A.D. (1994) Technology of cereals. Fourth edition. Pergamon Press.

Knight, C. and Stanley, R.P. (1999) Assured crop production: HACCP in agriculture and horticulture. CCFRA Guideline No. 10. Supplement 1: Lettuce case studies.

Leaper, S. (1997) HACCP: A practical guide. Technical Manual No. 38. 2nd Edition. Campden & Chorleywood Food Research Association.

MAFF (1987) Mycotoxins: Food Surveillance Paper No. 18.

MAFF (1989) Code of practice for the control of salmonellae in the production of final feed for livestock in premises producing over 10,000 tonnes per annum. MAFF Leaflet PB 0018.

MAFF (1993) Mycotoxins: third report. Food Surveillance Paper No. 36.

MAFF (1998) A review of antimicrobial resistance in the food chain.

MAFF (2000) Agriculture in the United Kingdom 1999.

Moore, D.M. (Ed) (1982) Green plant: the story of plant life on earth. pp 238-248. Domestication of plants. Cambridge University Press. ISBN: 0 521 24610 5.

Mortimore, S. and Wallace, C. (1998) HACCP: A practical approach. Aspen Publishers.

Noah, N.D., Bender, A.E., Reaidi, G.B. and Gilbert, R.J. (1980) Food poisoning from raw red kidney beans. British Medical Journal, 19 July: 236-237.

Pebody, R.G., Ryan, M.J. and Wall, P.G. (1997) Outbreaks of *Campylobacter* infection: rare events for a common pathogen. Communicable Disease Report Review **7** (3): R33-R37.

Phillips, C.A. (1995) Incidence, epidemiology and prevention of foodborne *Campylobacter* species. Trends in Food Science and Technology **6** (3): 83-87.

Ponzi, J. and Edwards, M.C. (2000) The effects of processing on foreign bodies. A case study with in-container heat processing. Review No. 16. Campden & Chorleywood Food Research Association.

Posner, E.S. and Hibbs, A.N. (1997) Wheat Flour Milling. American Association of Cereal Chemists.

POST (1997) Safer eating - microbiological food poisoning and its prevention. Parliamentary Office of Science and Technology. ISBN 1 897941 56 0.

Ranken, M.D., Kill, R.C., and Baker, C.G.J. (1997) Food Industries Manual. 24th Edition. Blackie Academic and Professional.

Ridgwell, J. (2000) Ready meals: video and workbook for schools. Ridgwell Press. ISBN: 1 901151 069.

Robinson, J.B.D. and Treherne, K.J. (1990) Maize. pp 41-49 In: Exploited Plants - collected papers from The Biologist. Institute of Biology ISBN 0 900490 25 X .

Roeder, R.A. (1991) Animal science and livestock production. pp 49-59. In: Encyclopaedia of Food Science and Technology (Ed J.H. Hen).

Rose, D. and Gaze, R.R. (1998) Safe packing of food and drink in glass containers: guidelines for good manufacturing practice. Guideline No. 18. Campden & Chorleywood Food Research Association.

Rose, D.J. (2000) Total quality management. In: Chilled Foods: A Comprehensive Guide (Eds: M.F. Stringer and C. Dennis). Woodhead Publishing, ISBN 1 85573 4990

Russell, P.L. (1982) Recent work on bread staling. FMBRA Bulletin (2): 69-80.

Scotter, C.N.G (Ed.) (1999) Food authenticity assurance: an introductory manual for the food industry. Guideline No. 23. Campden & Chorleywood Food Research Association.

Shaw, R. (1996) Product development guide for the food industry. Guideline No. 8. Campden & Chorleywood Food Research Association.

Shepherd, S. (2000) Pickled, potted and canned. The story of food preserving. Headline Press. ISBN: 0 747 223343.

Singleton, P. (1997) Bacteria in Biology, Biotechnology and Medicine. 4th Edition. John Wiley & Sons.

Singleton, P. and Sainsbury, D. (1978) Dictionary of Microbiology. John Wiley & Sons.

Smith, D.B., Roddick, J.G. and Jones, J.L. (1996) Potato glycoalkaloids: some unanswered questions. Trends in Food Science and Technology **7** (4): 126-131.

Stanier, R.Y., Adelberg, E.A. and Ingraham, J.L. (1976) General Microbiology. 4th Edition (MacMillan Press) p4.

Stringer, M. and Dennis, C. (2000) Chilled foods: a comprehensive guide. 2nd Edition. Woodhead Publishing/CRC. ISBN 1 855 73 499 0.

Stuart, J., Sufi, F., McNulty, C and Park, P. (1997) Outbreak of *Campylobacter enteritis* in a residential school associated with bird pecked bottle tops. Communicable Disease Report Review **7** (3): R33-R37.

Tamine, A.Y. and Robinson, R.K. (1999) Yoghurt: Science and Technology. Woodhead Publishing.

Thorne, S. (1986) The history of food preservation. Parthenon Publishing.

Thorpe, R.H. (1994) Guidelines on the prevention of visible can defects. Technical Manual No. 37. Campden & Chorleywood Food Research Association.

Timperley, D.A. and Timperley, A.W. (1993) Guidelines for the design and construction of floors for food production areas. Technical Manual No. 40. Campden & Chorleywood Food Research Association.

Timperley, A.W. (1994) Design and construction of walls, ceilings and services for food production areas. Technical Manual No. 44. Campden & Chorleywood Food Research Association.

Timperley, A.W. (1997) Hygienic design of liquid handling equipment. 2nd Edition. Technical Manual 17. Campden & Chorleywood Food Research Association.

Whitney, E.N., Cataldo, C.B. and Rolfes, S.R. (1998) Understanding Normal and Clinical Nutrition. Wadsworth Publishing.

ABOUT CCFRA

The Campden & Chorleywood Food Research Association (CCFRA) is the largest membership-based food and drink research centre in the world. It provides wide-ranging scientific, technical and information services to companies right across the food production chain - from growers and producers, through processors and manufacturers to retailers and caterers. In addition to its 1500 members (drawn from over 50 different countries), CCFRA serves non-member companies, industrial consortia, UK government departments, levy boards and the European Union.

The services provided range from field trials of crop varieties and evaluation of raw materials through product and process development to consumer and market research. There is significant emphasis on food safety (e.g. through HACCP), hygiene and prevention of contamination, food analysis (chemical, microbiological and sensory), factory and laboratory auditing, training and information provision. As part of the latter, CCFRA publishes a wide range of research reports, good manufacturing practice guides, reviews, videos, databases, software packages and alerting bulletins. These activities are under-pinned by fully-equipped modern food processing halls, product development facilities, extensive laboratories, a purpose-built training centre and a centralised information service.

In 1998 CCFRA established a wholly owned subsidiary in Hungary from where an experienced team of scientists and technologists provides training and consultancy on HACCP, quality management, product development, market and consumer research, food and environment law, and hygiene to Eastern Europe.

To find out more, visit the CCFRA website at www.campden.co.uk